# THE DEVELOPMENT OF TACTICAL SERVICES IN THE ARMY AIR FORCES

# THE DEVELOPMENT OF TACTICAL SERVICES IN THE ARMY AIR FORCES

*By John M. Coleman*

SUBMITTED IN PARTIAL FULFILLMENT OF
THE REQUIREMENTS FOR THE DEGREE OF
DOCTOR OF PHILOSOPHY IN THE FACULTY OF
POLITICAL SCIENCE, COLUMBIA UNIVERSITY

COLUMBIA UNIVERSITY PRESS
MORNINGSIDE HEIGHTS, NEW YORK
*1950*

Published in Great Britain, Canada, and India by
Geoffrey Cumberlege, Oxford University Press
London, Toronto, and Bombay

MANUFACTURED IN THE UNITED STATES OF AMERICA

*To Agnes*

## PREFACE

I have received invaluable help from many sources in preparing this study, but my greatest debt is to Captain (later Major) Franklin D. Walker, Chief of the Historical Office of the Air Service Command, who assigned me to this project and saw me through it. I am also indebted to all of my fellow historians at Wright Field, in both sections of the headquarters, for their kind cooperation and constant assistance.

Much of the information presented here was gathered from the interviews enumerated in the bibliography. The enlisted men, civilians, and officers, whose names I have included, gave me not only many facts, but a clearer understanding of the air forces, and I owe them much for their unfailing courtesy.

The historical office at Headquarters, AAF, in Washington, D.C., was helpful in securing the release of this material from classified status, and the history department of Columbia University was more than generous in accepting it as a Ph.D. dissertation. I am particularly grateful to Professor (now Dean) John A. Krout for sponsoring my work at Columbia, and for offering me much-needed guidance over a period of several years.

The following have given me useful criticisms which I have valued highly: Professor Henry Steele Commager of Columbia University, Professor Wesley Frank Craven of New York University, and Professor Joseph R. Strayer of Princeton University; also two of the leading protagonists: General Elmer E. Adler and General Joseph T. McNarney. Although I have had so much help, I take full responsibility for all statements of fact or opinion contained herein.

John M. Coleman

*Lafayette College*
*Easton, Pa.*
*September, 1949*

# CONTENTS

## FIGURES

# TABLES

★★★★★

# INTRODUCTION

THE PURPOSE OF THIS BOOK is to trace the development of the service and maintenance organizations of the Army Air Forces during World War II. The term "development" is interpreted to include the relevant aspects of manning, equipping, and training. The discussion begins with a brief account of the historical origins of AAF service and maintenance organizations, and considers in detail the factors affecting their wartime growth and subsequent decline. Emphasis is likewise placed on the manner in which military events and technological innovations influenced the organizational structure by which men and materiel were mobilized. In this sense the book is also a study of the dynamics of military bureaucracy.

The theme of the book is the application of broad organizational principles, both military and industrial, to the logistical problems of the air forces. For better or for worse, the chief trend is toward centralization. At first, as one of the generals in the program remarked, "There wasn't sufficient facility . . . in the war effort to know exactly where the theaters [of operations] were to be." [1] Nor was there an adequate plan for the support of air combat units. So much was going on, and in such haste, that no one seemed to know quite what was happening or how to find out. The problem of the higher headquarters was to obtain reliable information, and if possible to assume control over the events for which they might be held responsible. Success in obtaining such control was achieved only by centralizing the channels of information and authority.

As a matter of present concern the condition of the armed forces in the early stages of World War II should not be forgotten. This book documents the complete unpreparedness for war which characterized all maintenance and supply agencies responsible for Air Corps logistics. The confusion which the Japanese attack on Pearl Harbor produced in all elements of command, from the Washington headquarters to the lowliest squadron orderly room, is persuasive evidence of the dream-like quality of American preparations for total war.

Shortages and bottlenecks in the support of air power explain in part not only the failure to protect the Philippines and the Dutch East Indies in 1941, but also the gamble that was taken in the decision to invade Europe in 1944. Plans for a tremendous expansion of military strength in England in the summer and fall of 1943 were based on a tentative D-Day, sometime in the following spring. In the face of a continuing threat from the German Air Force, it was decided that air force units would have to be shipped first. On the one hand, aerial attack on Axis defenses was regarded as an indispensable preliminary to invasion; on the other, it was estimated that virtually all shipping after the first of the year would be required for the assault forces. And in this situation it was found that the AAF was not prepared to ship the needed service units.

The significance of the administrative arrangements described in this study, such as the McNarney Directive, the Bradley Plan, and General Adler's training structure, can be seen in the light of these great events. The story is here told from the point of view of the Wright-Patterson Field headquarters, near Dayton, Ohio, a command with a primary responsibility. Although not a battle narrative, this history may cast light on one aspect of air power which cannot be ignored in any battle.

Unfortunately, it is impossible to write about military aviation without using army language. And army language, in addition to being technical and wordy, is also confusing, for its meaning is frequently less clear than it sounds. For example, what civilian would know that *organization* is a technical term applying to military forces only at the regimental or group level, and that it must be distinguished sharply from *unit?* The distinction must nevertheless be made, for this book discusses only group-level organizations, and those units which operated with them. It should also be made clear that such separate services as the component parts of the Weather Wing are not discussed here, for these units were usually deployed as regional squadrons.

In a study of this sort there is no escape from the language difficulty; translations only make matters worse. Army language can at least be understood by those who take the trouble to learn it. Translations, in matters of detail, can be understood by nobody. Yet there is one aspect of army language for which an apology must certainly be made—the constant changing of names and terms. Not only units and organizations, but even the higher headquarters, were regularly redesignated throughout the war, and this process had apparently been going on for a long time. People who were temporarily connected with the army often wondered whether these everlasting redesignations were intended to confuse the enemy, to enrich the manufacturers of office stationery, or perhaps to satisfy the creative instincts of newly assigned commanding officers.

To help the reader as much as possible, many terms are explained in the text or in footnotes, and there is a long glossary of abbreviations at the end of the book. References in the footnotes or elsewhere to the Supporting Documents of this study, or of other studies, indicate collections of materials which are now on file in Washington, illustrating the original monographs.

This work particularly emphasizes Air Depot Groups and Service Groups, which were trained by the Maintenance Command and its successors, the Air Service Command and the Air Technical Service Command. Consideration is also given to precursor units, such as the Service Companies of the Corps Areas and the Air Base Groups of the old General Headquarters Air Force, because a knowledge of these organizations is essential for an understanding of Air Depot Groups and Service Groups. The discussion is not limited to Air Corps units, but includes the units of the attached arms and services of the Army Service Forces.* Finally, from the units themselves the study turns to the persistent efforts of the central command to establish and perfect a unified national training structure.

The pattern of development was early influenced by the military situation. When the United States entered the war, superiority in the air belonged to the Axis powers both in Europe and in Asia. Of necessity the combat units of the air forces began to operate from dispersed squadron airdromes. Each independent organization was expected to perform its own internal administration and "housekeeping." Supply and maintenance functions, formerly performed by base service units,

* The term *arm* applies to combat branches of the army, such as the Infantry or the Artillery. The term *service* applies to non-combat branches, such as the Quartermaster Corps or the Finance Department. The classification is not exact, as most branches have units of both types.

were taken over by fighter groups and bomber groups. Thus, personnel and equipment were added to combat organizations—with disastrous military results. Instead of being streamlined for action, these units were made more cumbersome and unwieldy. Thus, early in 1942, it was necessary for General Arnold to direct a complete reorganization of all AAF units to effect wherever possible a 35 percent reduction of personnel and equipment.[2] The logistical system resulting from this reorganization eventually became the basis of American air power throughout the world.

The concept of one Service Group for two combat groups and one Air Depot Group for two Service Groups, which was set forth in AAF Regulation No. 65-1, 14 August 1942, was evolved during General Arnold's reorganization. In turn, this concept resulted in a delineation of functional responsibilities, as defined in four echelons of maintenance and supply, and in the decision that first and second echelon work would be the responsibility of the combat air forces, and third and fourth the responsibility of the Air Service Command (as it was then called). The distinction is significant here because it is with the activities of the Air Service Command that this study is primarily concerned.

There was nothing new, of course, in the idea of echelons of maintenance; automotive maintenance and mechanized ground activities had been organized in that manner for years. Indeed, it was only natural that the army should apply the technique with which it was familiar to the similar problems of military aviation. But the assignment of third and fourth echelon work to a separate command, responsible for the establishment of depots and sub-depots and the activation and training of Air Depot Groups and Service Groups, was a decision of the first importance in the history of the Army Air Forces. Although some third and fourth echelon service units were to remain under commands other than the Air Service Command, there were comparatively few exceptions to the general rule.

First echelon aircraft maintenance was defined in AAF Regulation No. 65-1 as that performed by air elements of combat units; second, that performed by ground elements of combat units, by air base squadrons, or by airways detachments; third, that performed by Service Groups and sub-depots; and fourth, that performed by Air Depot Groups and air depots. Echelons of maintenance in the other arms and services were described respectively as maintenance performed by the equipment operators, the using organizations, the service center organizations, and the depot organizations. Echelons of supply were placed on the same basis, but included the additional connotation of time: 3-day level of supplies, 10-day level, 30-day level, and 90- to 150-day levels.[3]

Traditionally, many third and fourth echelon services to the air forces had been divided between the Army Air Forces and the Army Service Forces. The training of service and maintenance organizations, therefore, had to be broken down into two main categories: the training of Air Corps units for aircraft services and the training of ASWAAF units (arms and services with the Army Air Forces) for associated services. As the war progressed, an increasingly large proportion of the load was borne by the air arm, and ultimately the other arms and services were almost completely integrated into the structure of the Army Air Forces. But during 1942 and 1943, when the bulk of these units were prepared, much of the preliminary organization and training of ASWAAF's was accomplished in fact by the respective branches of the Army Service Forces and by the domestic air forces, as well as by the Air Service Command. In other words, the Air Service Command was responsible for coordinating and influencing many activities over which it had no direct control, in order to supervise the preparation of units which would eventually come under it. The problem of creating a separate air force, which has been the subject of so much controversy, should perhaps be considered in the light of the administrative and logistical difficulties described in this book. No one was more aware than the people who had to develop service units that if the air forces were suddenly to become separate, they could hardly expect to live in the manner to which they had become accustomed.

★★★ 1 ★★★

# ORGANIZATIONAL AND TACTICAL BACKGROUND

## Brief History of the Army Air Arm

The army air forces of World War II may be said to have originated in 1907 as the Aeronautical Division of the Signal Corps.[*] The pioneer airmen of that period immediately asked for separation from the Signal Corps,[1] since that branch could assign to aircraft only the limited duties of auxiliary reconnaissance and communications.[2] They were convinced that aviation was destined for a unique place in military strategy and that airplanes were more than just "the eyes of the army." Their ambitions were not satisfied overnight, but in 1914 the division was promoted by the Signal Corps to the status of Aviation Section.[3] During World War I this section was abolished, and instead there was created under the War Department a Division of Military Aeronautics, while in France the air component of the American Expeditionary Force was organized as a tactical Air Service.[4] Toward the end of the war it was realized that air forces were capable of making heavy attacks against the enemy far behind the lines of surface combat.

By the terms of the Army Reorganization Act of 4 June 1920, an independent Air Service was created for the army as a whole; [5] and as a full-fledged combat branch of the line the new Air Service received appropriations and assumed responsibility for the maintenance and supply of its own technical equipment.[6] In 1926 this organization was redesignated the Air Corps. At the same time, in recognition of its increasing importance, it was accorded sectional representation in the War Department General Staff.[7] Until the activation of the General Headquarters (GHQ) Air Force in 1935,[8] the Air Corps carried on all army aviation

[*] There had been earlier aeronautical activities, such as the Balloon Corps of the Civil War and the Balloon Detachment of the Spanish-American War, but these had had only lighter-than-air craft. They were hardly the direct ancestors of the organizations with which this study is concerned.

activities and was the sole branch of the army air arm. After 1935 the Air Corps and the GHQ Air Force existed side by side. The logistical difficulties of this arrangement will be discussed later; the point to be made here is that it was not until 1941 that these two branches were united to form the Army Air Forces,[9] and even then the AAF was only one of about twenty separate commands, each supposedly entitled to direct access to the Chief of Staff. One of the first things required when the United States entered the war was a sweeping reorganization of the War Department.

Just before the Japanese attack in the Pacific, General George C. Marshall, Chief of Staff, had called Brigadier General Joseph T. McNarney back from England and had delegated to him the task of supervising this reorganization.[10] General McNarney left England on 7 December 1941; before he could begin his new assignment he had to fly to the Hawaiian Islands to participate in the investigation by the Roberts Committee of the damage done at Pearl Harbor. Immediately after his return, he and a group of representatives from each of the branches of the army set to work at top speed and in great secrecy. The result of their labors was the consolidation of the War Department on 9 March 1942 under three coordinate, autonomous commands: the Army Air Forces, the Army Ground Forces, and the Services of Supply (later redesignated the Army Service Forces).[11] Simultaneously, the General Staff was reorganized so that approximately 50 percent of its personnel would be representatives of the air arm.

This reorganization was a great promotion for the Army Air Forces, or rather a great demotion for the other branches, which were all placed under the three top headquarters.[12] Indeed, the Inspector General's Department was the only independent command to retain its former status. Since the function of the Inspector General was to check up on all other branches, including the air forces, it was deemed advisable by the McNarney Committee to allow this one branch to remain independent and to retain direct access to the Chief of Staff.

Thereafter, until the post-war consolidation of the army, the navy, and the air force into a single Department of Defense, the position of the AAF within the army remained generally the same. In other words, the foregoing description is applicable to the whole war period, except for the first few months, and it was with this triple structure, represented by the AAF, the AGF, and the ASF, that the public became familiar. It should be noted, therefore, that this form of organization was achieved only after the United States had entered the war. In the pre-war period, now to be considered, the air forces had inadequate representation on the General Staff, they were split into two competing services, and these services together constituted only one of about twenty in the War Department.

### The Air Corps and the GHQ Air Force

From 1935 until 1941 responsibility for the aerial activities of the United States Army was divided between the Air Corps and the GHQ Air Force (later redesignated the Air Force Combat Command). This responsibility was shared with the nine Corps Areas (later redesignated Service Commands), which were the basic geographical subdivisions of the army, and were in no way responsible for the combat effectiveness of the air arm. Theoretically, the Air Corps was charged with administrative functions, and the GHQ Air Force with operational functions, while the Corps Areas "assisted" both. But as a result of divided responsibility, an inadequate system of supply and maintenance was developed, and neither of the two main branches of the air arm had undisputed authority to train service units for the several related commands.

Prior to 1935 tactical units had been under the nine Corps Areas. Supply and maintenance had been performed by Service Companies, which were indicated on organizational charts as component units of combat groups. In practice, Service Companies (later Service Squadrons) were often stationed during the 1920's and early 1930's at bases where there were no groups at all, for the purpose of manning base headquarters.[13] In a few cases, on the other hand, two or more groups were stationed at the same base.[14] The maintenance functions of the respective Service Companies were combined, and the joint activity functioned under the supervision of the base commander, who was required by regulations [15] to designate the senior engineering officer of the station as the "Chief Engineer Officer." * The mission of Service Companies, in any case, was to perform emergency maintenance operations "the accomplishment of which by personnel of tactical units would detract from the tactical efficiency of such units, and the magnitude of which would not

* Later the term *engineering*, rather than engineer, was used for aircraft maintenance functions, to avoid confusion with the functions of the Corps of Engineers.

warrant withdrawal of the equipment from the group or station for depot overhaul." [16]

From 1922 until 1930 airplanes were sent to the depots only when they were badly damaged or in need of major overhaul, as indicated by periodic inspections. Airplanes and engines from World War I were still plentiful: a policy of economy had not yet been introduced.[17] After 1930, when the overhaul system was standardized,[18] army airplanes were automatically sent in for depot overhaul at predetermined intervals. This arrangement was more economical and at the same time more effective, for a closer supervision was exercised by the depots over the servicing of all army planes.

When the GHQ Air Force was activated on 1 March 1935, the Commanding General thereof was given authority to direct air force organization and training, including maneuvers, and to supervise the inspection, maintenance, and operation of technical equipment. He was to be responsible to the Chief of Staff in peacetime, and to the Theater Commander in time of war.[19] The creation of the GHQ Air Force was an important step in granting independence to the air arm, for the new organization was given complete control over tactical units. The nine Corps Areas, however, retained administrative jurisdiction over bases at which tactical units were stationed [20] and exercised their authority through the non-mobile engineering, supply, and transportation departments of the local Air Base Squadrons. Moreover, the Corps Areas retained courtmartial jurisdiction over Air Corps * personnel.[21] As yet the independence of the air arm was more apparent than real.

Expansion in the form of construction was likewise under the supervision of the Corps Areas. Construction for the army during the 1920's and 1930's had been the responsibility of the Quartermaster General. On 1 January 1941, however, construction at air bases was transferred from the Quartermaster General to the Chief of Engineers,[22] and a similar change was subsequently made for the rest of the army. Since both the Quartermaster and the Engineers operated through the Corps Areas, construction at air bases remained under Corps Area control.

Responsibility for Air Corps supply and technical control over maintenance were functions of the Chief of the Air Corps. During the latter

* Air arm personnel were called Air Corps personnel, whether they were part of the GHQ Air Force or of the Air Corps. The term Air Corps, as applied to personnel, was continued throughout World War II.

part of the 1930's the Chief of the Air Corps, through the Materiel Division and its Field Service Section at Wright Field, Ohio, also supervised fourth echelon maintenance and all Air Corps supply at the four continental air depots, which serviced both Air Corps and GHQ Air Force units.[23]

At the bases where most of the squadrons and groups were located, maintenance and supply activities, other than third and fourth echelon, were performed by base engineering and supply departments, which were under the administrative jurisdiction of the Corps Areas. Only "exempted stations" [24] were excluded from the administrative control of the Corps Areas, and assigned directly to the chief of one of the two branches of the air arm. Even at "exempted stations," however, tactical combat organizations were often serviced by the base engineering and supply departments.

Combat groups in peacetime requisitioned and stored their own spare parts and equipment.[25] The procedure they followed was an interesting revelation of the true situation of air force organizations. When needed items were not on hand, Service Companies (or Service Squadrons) assigned to the combat groups requisitioned from base supply officers before going to the four air depots. Requests were thus channeled through installations officially subordinate to the nine Corps Areas, thereby subjecting air force activities to Corps Area inspections and audits. When this formality had been completed, base supply officers—or in some cases the Service Company supply officers themselves—requisitioned from control depots, which filled the requisitions if possible, and extracted (that is, forwarded requests for) unavailable items to the Field Service Section at Wright Field, Ohio.[26] Not until the establishment in the fall of 1940 of four Air Districts (later to become the four continental air forces) were air arm troops at GHQ Air Force stations removed from the surveillance of Corps Area commanders.

A new tactical organization was also introduced in 1940, when the GHQ Air Force established the Air Base Group.[27] Normally this organization consisted of a Headquarters and Headquarters (Hq and Hq) Squadron, a Materiel Squadron, an Air Base Squadron, and several attached units from the other arms and services. Its purpose was to perform first and second echelon maintenance and supply for one combat group, but, reinforced by an additional Materiel Squadron, it could service two combat groups. In drawing up plans for this group, the G-4

(Supply) Section of the GHQ Air Force started from scratch, "for we were unable to find data on any country describing a well-planned air force logistical system which could be used as a guide." [28] In general, the services it performed were limited to minor repairs and "housekeeping." Although third echelon maintenance was mentioned in some of the directives, what later became third echelon work was not attempted by any of the first Air Base Groups. Nor were they mobile in the sense of accompanying combat groups from base to base.

Difficulties in the administrative technicalities of supply and maintenance, even with the few airplanes available in peacetime, revealed inadequacies in the system. Air Corps officers, most of them flying personnel, were losing valuable time in the air attending to ground duties and routine administration.[29] Even more serious was the loss of mobility in combat units; tactical units of all kinds were acquiring more personnel, more equipment, and incidentally the need for more shipping space.

### The Maintenance Command of the Air Corps

In 1941 an entirely new element of command was introduced into air force logistical organization with the establishment on the 15th of March of the Provisional Air Corps Maintenance Command as a part of the Materiel Division of the Air Corps.[30] This organization was headed by Colonel Henry J. F. Miller, with Lieutenant Colonel Elmer E. Adler as Chief of Staff. On the 24th of April the word "Provisional" was dropped, as the new command was considered to have passed its trial period.[31] The establishment of this command followed the withdrawal of Corps Area jurisdiction over the administration of air bases in 1940, and was the result of the prospective increase of Air Corps training facilities.

In the beginning the Maintenance Command consisted of a Headquarters, the Field Service Section, the 50th Transport Wing, and six air depots, which were situated at Fairfield, Ohio; San Antonio, Texas; Sacramento, California; Middletown, Pennsylvania; Ogden, Utah; and Mobile, Alabama. In addition, there was established the Fairfield Maintenance Group Area and the San Antonio Maintenance Group Area, each consisting of a control depot and several well-established Air Corps sub-depots.[32] At the bases where these designated sub-depots were located, station engineering and supply activities passed to the control of the Maintenance Command at the time of its activation. Existing relationships between GHQ Air Force stations and the control depots were not altered until January, 1942.[33]

Thus, the Maintenance Command system at first applied only to a few bases under the jurisdiction of the Chief of Air Corps. For several months the GHQ Air Force (Air Force Combat Command) continued to perform its own first, second, and occasionally third echelon work at engineering, supply, and transportation establishments at GHQ Air Force stations. It was during this period, early in 1941, that Air Base Groups were organized by the GHQ Air Force at a number of bases in the continental air forces.[34] Four of these groups at Mitchel, Selfridge, Orlando, and Savannah, under the leadership respectively of Colonels Paul E. Ruestow, Edgar T. Selzer, Robert Easton, and Bernard Castor, developed rapidly during their experimental stages and became the forerunners of the more recent Service Groups.[35]

Meanwhile, the Maintenance Command activated four new mobile units, called Air Depot Groups, to accompany combat units overseas and to perform second and third echelon functions. And on 28 August 1941 the Maintenance Command began the installation of civilian sub-depots, with commissioned personnel in charge, to perform third echelon maintenance and supply at *all* domestic stations under the direct control of the Chief of Air Corps.[36] The presumption was that sooner or later sub-depots would be established throughout the air forces.

The GHQ Air Force did not concur in plans to establish sub-depots at the bases of the numbered air forces. They protested that sub-depots were neither mobile, nor in any sense of the word field organizations appropriate to a combat command, and succeeded in postponing the establishment of sub-depots at GHQ Air Force (Air Force Combat Command) bases until January 1942.[37] As an alternative to the sub-depot program, they continued the development of the Air Base Group. Yet in spite of all that was done to make this organization a success, base commanders were reluctant to push it, being afraid that their efforts would be nullified if the sub-depot became a reality.[38]

Tactical service units of a sort were thus developed by both main branches of the army air arm: Air Base Groups by the GHQ Air Force and Air Depot Groups by the Maintenance Command of the Materiel Division of the Air Corps. The need for mobility was recognized; without mobile service units this country would be unable to extend its air power to the fighting fronts. But until the jurisdictional conflict between the two branches could be resolved, there was little likelihood that a single, comprehensive logistical system would meet universal agreement. The installation of the controversial sub-depots in August 1941 and January 1942, and consequent hesitation and confusion on the part of the base commanders, temporarily checked the training of Air Base Groups, with the effect that these organizations never presented more than a partial solution for the problems of overseas air forces.

### The Army Air Forces

On 20 June 1941 the long overdue merger of the Air Corps with the Air Force Combat Command (formerly the GHQ Air Force), to form the Army Air Forces, was finally consummated.[39] By 17 October 1941, when the Air Service Command replaced the Maintenance Command [40] and assumed responsibility for third and fourth echelon maintenance and supply for the Army Air Forces, the process of consolidation was nearing completion. On 11 December 1941 the new service command was released from the jurisdiction of the Materiel Division and assigned directly to the Chief of Air Corps.[41]

Part of the reorganization in Washington was the transfer of the G-4 (Supply) Section and most of the Special Staff of the old GHQ Air Force to Headquarters, Army Air Forces, where they continued to function under the name of the Directorate of Base Services.[42] With them they brought the logistical plans developed under their former command. On 9 March 1942, Colonel Lyman P. Whitten, formerly of the Office of the Chief of Air Corps, became Director of Base Services, and assumed responsibility for coordinating at the Washington headquarters the equipping of all AAF service organizations.

At the time Colonel Whitten took over, several important changes remained to be made. One of these was the transfer of Air Base Groups (simultaneously redesignated Service Groups) to the Air Service Command—which was not finally accomplished until 22 June 1942.[43] Another was the establishment of Station Complements at various bases in the Air Service Command, a move which was necessitated by the transfer of Quartermaster, Signal, Ordnance, Chemical Warfare, Finance, Medical, and other services from the base commanders to the Service Groups. The latter development occurred in point of time both before and after the transfer of arms and services personnel to the Service Groups of the Air Service Command, but it was largely the result of that transfer.[44] The first eleven Station Complements were activated in April 1942; others were activated in May and June.[45]

By the time the component units of Air Base Groups (Service Groups) were turned over to the Air Service Command, 31 of these groups, and 11 Air Depot Groups, had been shipped overseas.[46] Thereafter Service Groups remaining in this country were trained by the Air Service Command to perform in the theaters of operations the same third echelon work as that accomplished in this country by sub-depots. Air Depot Groups, ultimately, assumed responsibility for fourth echelon functions, which had been performed in continental United States only by air depots. As the characteristic organizations of the Air Service Command, Air Depot Groups will be discussed first. Service Groups, which subsequently became much more important, will be brought into the story at the time they were reorganized, June 1942. Before that, however, it will be necessary to consider the development of overseas air forces, with especial reference to the organization of service functions, and the tactical situation which created the need for mobile service units.

## Plans for Overseas Air Forces

The organization of the army air arm with which this country entered the war was largely established and developed by staff elements of the old GHQ Air Force. After its reorganization in 1940 the GHQ Air Force (redesignated Air Force Combat Command early in 1941) comprised a headquarters and the First, Second, Third, and Fourth Air Forces, each of which was organized with a high degree of flexibility to facilitate the reinforcement of departmental or overseas air forces, and the creation of task forces for specific operations. These four air forces continued to exist after the merger of the Air Force Combat Command with the Air Corps, and two of them, the First Air Force and the Fourth Air Force, assumed additional functions in support of the Eastern Defense Command and the Western Defense Command of continental United States.[47]

At first, each of the overseas air forces was organized in a manner peculiar to the conditions of the theater in which it operated. Some had Service Commands, and some did not. On 14 May 1942, however, the general form of these air forces was standardized so far as supply and maintenance were concerned. A directive of that date stated that "there will be established in each air task force, designated as an overseas air task force, a service command to service and function under the direct command and supervision of the Commanding General of the air task force concerned." [48] The commanders of Air Task Force Service Commands were advised that they could organize the agencies, facilities, and military units under their control in such manner as they deemed necessary.

The reasons for the establishment of two types of logistical organization, one for the United States and one for overseas, and the differences between the two, were clearly explained at the beginning of AAF Regulation No. 65-1, 14 August 1942, in a descriptive passage entitled "Continental U.S. vs Task Force":

Due to differences in service functions required of task forces as against operations within the continental limits, it is necessary to establish two plans. Within the continental limits, service functions are increased to include the activation, organization, training and equipping of service units to make up task forces. Due to the fixed situation, more extensive permanent construction will be undertaken, and more elaborate facilities for supply and maintenance established. Within the continental limits, zone of the interior procurement and supply agencies must be provided to secure and ship materiel to all task forces as well as all air forces and training activities therein. For this reason, *responsibility for the most service functions at home rests with a command separate from the air forces, namely the Air Service Command. In an air task force, service functions revert to control of the air force commander.*[*]

It is interesting to note, however, that one of the first suggestions for a service command of any sort arose in connection with the Second Air Force. This suggestion was made in 1941, before the command principle had become fully established in continental United States, and the task force principle in overseas air forces. It came from Colonel Harold A. McGinnis, who was at that time Air Force Inspector General, but was destined to go to the Eighth Air Force Service Command in England, the largest and most important of the early overseas organizations of this type. He recommended in an inspection report of 25 July 1941 that a command be set up for the Second Air Force to supervise Air Corps supply and maintenance, and all common supplies and services, including Quartermaster, Ordnance, Chemical, Signal, Medical, and Finance. It was his desire that

a Service Command be organized on a line of authority coequal to that of the Interceptor Command, Bomber Command, and the Air Support Com-

[*] Italics not in the original.

mand, as a Command Agency under the 2nd Air Force, to coordinate and, in the name of the Air Force Commander, administer the general administrative and service functions of all Air Bases of the Command.[49]

Each of the Air Forces in the United States considered a similar recommendation in drafting plans for expansion and reorganization during this period. But separate service commands were never set up in the domestic Air Forces. Instead, as has been pointed out, there was established an over-all Air Service Command under the Chief of the Air Corps to coordinate supply and service for all elements of the Army Air Forces. Colonel McGinnis's recommendation, therefore, was not acted upon until a command of the sort he desired was set up in the European theater.

Whenever possible, pre-war plans for Air Force Service Command operations were based on standard procedures. This policy was most noticeable in the detailed arrangements for Air Depot Groups and Air Base Groups. Provision was made in these arrangements for the pooling of bulk supplies, both Air Corps and other arms and services, at certain centrally located bases, in accordance with all the established army and air force procedures.

In view of the scope of the supply problem, and the thousands of individual items required for the maintenance of military aircraft, it was anticipated that large installations would be required. Thus, in the event of war, a few air bases overseas would become in fact air force general depots (comparable to air depots at home, or army general depots in the theater). Moreover, plans drawn up by the GHQ Air Force early in 1941 stated that "more than one such air base may be required in certain theaters where great distances are involved." [50] Participation by the other arms and services in supplying the air forces was to be dependent in each case upon the availability of army, communications zone, and zone of the interior supply points suitably located. Whenever ground force or fixed Air Corps supply installations were available, they were to be utilized by the air forces.

Obviously, these plans were not very specific. Emphasis on defense was characteristic of American thinking at that time—military as well as civilian. The expansion of the entire army, according to General Marshall, was based on "the long-standing protective mobilization plan," [51] developed in peacetime before the necessity for crossing oceans and suppressing aggressor nations had materialized. Not a single AAF unit had been service-tested in combat (as Japanese units had been tested in China, Italian units in Ethiopia, and German units in Spain). In fact, only a few American units had experienced field conditions on maneuvers, and this experiment, late in 1941, had taken place under such abnormal conditions that it was impossible to draw from it any but tentative conclusions.

### The Need for Tactical Service Units

Difficulty in deploying our forces in the far eastern area at the beginning of the war demonstrated clearly the consequences of national unpreparedness and failure on the part of the air forces to develop mobile supply, maintenance, and repair organizations for overseas service.[52] The loss of Wake Island and the isolation of the Philippines had necessitated the creation in January 1942 of an alternate trans-Pacific route via Christmas Island, Canton Island, Fiji, and New Caledonia.[53] While this route was being secured, air movements by heavy bombers were undertaken from Miami, Florida, through Brazil, equatorial Africa, India, and Sumatra to Java and Australia.[54] Before the termination of deliveries by this route, following the loss of Sumatra in February 1942, several spectacular flights more than halfway around the world were made to reinforce our besieged garrisons in the Pacific.

Typical of the experience of these valiant forces was the story of the 9th Squadron (Replacement), which flew across the Atlantic, North Africa, India and Burma to Java, carrying crews and a few extra men, but no bulky supplies or equipment.[55] When this squadron arrived in Java, its twelve B-17's landed at dawn, having flown on the last leg of the flight 2,600 miles from Bangalore, India. That very afternoon they were ordered to bomb up, and the following morning the first flight of three was sent out on a bombing mission. These were the first Flying Fortresses with tail guns to reach that theater. For a short time their success was remarkable. Japanese fighters, familiar with earlier models of B-17's were in the habit of coming in from behind with comparative safety. This particular flight shot down at least six Zekes before the Japanese became aware of its increased armament. Other flights too enjoyed a brief success.

Expeditions of this sort, however, could offer only a forlorn hope to the Allies. Although the planes had been flown to the scene of action in less than ten days, the ground units and materiel to keep them flying in the theater would have required two and a half months or longer for the transfer. There were no spares to replace losses, and few supplies; repair was difficult without mobile depot facilities; and what equipment there was could not readily be moved. Consequently, these Fortresses, flying one, two, or three at a time, without adequate support on the ground, never had a chance to develop a real offensive. Orders came to hold on as long as possible, then to take everything that could fly to Australia. When these war-weary ships finally pulled out of the Dutch East Indies, there wasn't much of an air force left. Improvised expedients had failed to satisfy the logistical necessities of modern warfare. The strategical lessons involved in this and similar ventures were summarized by General Marshall as follows:

> While this sudden reversal of a movement halfway around the earth demonstrated the mobility of the airplane, it also demonstrated the lack of mobility of air forces until a lengthy process of building up ground service forces and supplies (mechanics, ordnance and radio technicians, signal personnel, radar warning detachments, antiaircraft, medical and quartermaster units, as well as the troops to capture airfields and defend them against land attack, and the accumulation of repair machinery, gasoline, bombs, and ammunition) had been laboriously completed by transport plane, passenger and cargo ship—the last two largely being slow-moving means of transportation.[56]

★★★ 2 ★★★

# FIVE AIR DEPOT GROUPS IN 1941

## Activation

ALTHOUGH plans for meeting the service requirements of the air forces had been made in the late 1930's, there were no mobile service units in either branch of the army air arm at the close of the year 1940. The need for these units, however, had been clear to at least a few air force officers long before the great disaster in the Pacific.[1] Preliminary steps had therefore been taken for the constitution by the War Department of paper organizations to be used as mobile units whenever required. The new units were listed in the War Department Protective Mobilization Plan for 1940 as Regular Army, Inactive, Headquarters and Headquarters (Hq and Hq) Squadrons, Air Depot; Supply Squadrons; and Repair Squadrons.

When the need became acute, these units could presumably be called into being and dispatched either to the scene of conflict or to reinforce one of the departmental base depots, which had been established in Panama, Hawaii, and the Philippines in 1927.[2] But none of the new units had been activated, manned, or tested prior to 1941, and the plans on which they were based were not specific as to sources of personnel or methods of training. In September 1941, and later, consideration was still being given to the establishment of overseas depots "similar to Air Corps Mobile Repair Depots . . . but to be operated by civilians throughout." [3]

By this time the importance of intensified preparations was becoming obvious. It was evident, for example, that the British in the Near East and in England could perform only first and second echelon work on American airplanes with the tools and spare parts obtained through "Cash and Carry." [4] Third and fourth echelon work, if done at all, would require American equipment and personnel. The idea of furnishing civilians for this task was perhaps a natural one, but too optimistic when considered in the light of the difficulties experienced by such commands as the Panama Canal Department. It was here indeed that the new Air Depot Group program was begun.

The 1st Air Depot Group was activated at France Field, Canal Zone, effective 1 January 1941.[5] Its purpose was to provide the Panama Air Depot with a body of enlisted men to assist in the expansion of defense facilities at the Panama Canal.[6] The reason for its activation was the fact, already indicated, that departmental authorities had been unable to obtain a sufficient number of civilians, and new hands were needed as mechanics, clerks, stock tracers, truck drivers, and for a host of other interrelated jobs connected with a primarily civilian air depot. The 1st Air Depot Group was thus neither a mobile nor a tactical unit, and it was never trained in the orthodox manner—until its transfer to Kelly Field, Texas, in June 1944. For these reasons it was something of a special case, and not in any way an effort to solve a general logistical problem.

The 2nd, 3rd, 4th, and 5th Air Depot Groups, on the other hand, were experimental, and of much greater importance in the development of wartime logistics. Late in the year 1940, Brigadier General George H. Brett, Acting Chief of the Air Corps, proposed that four Air Depot Groups be activated, one at each of the existing air depots, to determine the suitability of tentative Tables of Organization (T/O's), Tables of Allowances (T/A's), and Tables of Basic Allotments (T/BA's).[7] He further proposed that one of these groups be equipped and trained for arctic service, one for tropical or jungle service, and two for service in a temperate climate.[8] To execute this plan, General Brett said, it would be necessary for 2,000 enlisted men to be recruited and allotted to the Air Corps. These men were required in addition to the troops already available to support the "54 Group Program," which was a plan based on the procurement of 12,835 airplanes by 1 April 1942, approved in June 1940 as the army's first aviation objective for training, organization, and procurement.[9]

Anticipating approval of this proposal, the Buildings and Grounds Office in Washington requested in December 1940 that funds amounting to $1,600,000 be included in supplemental estimates for the fiscal year 1941 for construction of housing and technical buildings at various places for the proposed groups.[10] It was observed that each of the four air depots was already overcrowded. Before additional troops could be ordered to these fields, temporary barracks, mess halls, and recreation buildings would have to be provided.[11] Meanwhile, the groups could be quartered in tents and trained in existing warehouses and hangars. This request was supported by another in April 1941:

There is an immediate need for facilities with which to train approximately four depot repair wings for mobile depot repair services under the new Maintenance Command organization. Two thousand enlisted men will be assigned immediately to four depot repair groups, and no facilities exist at the depots to which they are assigned for training purposes. Immediately following the activation of these four groups, additional groups will be activated and immediately placed in training at the existing depots, and as new depots open, repair groups will be placed in training thereat. These hangars will be utilized as training buildings, and will be used jointly by the military repair depots and the civilian employees in pre-work and upgrade training. . . . No facilities now exist to train these civilians.[12]

Before construction was approved, however, The Adjutant General of the War Department issued an order giving the final decision as to the activation of the new groups.[13] The Commanding Officers of McClellan Field, California, Duncan Field, Texas, and Patterson Field, Ohio, were instructed that four organizations, listed below, had been constituted and would be activated * on or about 1 April 1941 from Air Corps personnel at the stations concerned. Locations would be as indicated in Table 1.[14]

Within four days the groups were activated, and training began forthwith. The 2nd Air Depot Group, for example, was activated at Sacramento[15] on 1 April with four officers and eighty-three enlisted men. Training in the depot engineering and supply departments began on the following morning, the assignment of soldiers being staggered to enable the departments of the depot to absorb additional personnel without undue disturbance. But the mission of the new unit was not clear; on 10 April 1941 Lieutenant Colonel A. G. Liggett, Depot Supply Officer at Sacramento Air Depot, wrote to Colonel Henry J. F. Miller, Commanding Officer of the Provisional Air Corps Maintenance Command, indicating his uncertainty as to the mission of Air Depot groups:

In addition to my Depot Supply job, etc., I have drawn the assignment of Commanding Officer of the Second Air Depot Group, and would like to know just what this group is set up to do and to what extent. . . .

* For a discussion of the terms *constituted* and *activated*, see p. 121.

TABLE 1

ACTIVATION OF THE FIRST FOUR AIR DEPOT GROUPS

| *Unit* | *Station of Activation* | *Permanent Station* |
|---|---|---|
| 2nd Air Depot Group<br>Hq & Hq Sq<br>2nd Repair Sq<br>2nd Supply Sq | McClellan Field, Sacramento, Calif. | McClellan Field |
| 3rd Air Depot Group<br>Hq & Hq Sq<br>3rd Repair Sq<br>3rd Supply Sq | Duncan Field, San Antonio, Texas | Duncan Field |
| 4th Air Depot Group<br>Hq & Hq Sq<br>4th Repair Sq<br>4th Supply Sq | Patterson Field, Fairfield, Ohio | Patterson Field |
| 5th Air Depot Group<br>Hq & Hq Sq<br>5th Repair Sq<br>5th Supply Sq | Duncan Field, San Antonio, Texas | Mobile, Ala. |

There are many matters I should like to discuss with you, and others at the Division, but have been unable to get away. How about requesting orders for me for two or three days' visit at the Division? [16]

Colonel Liggett was also anxious to know what the probable assignment of the group would be; in particular, whether the group should train for tropical, arctic, or temperate climate. In another letter, addressed to the Assistant Chief of the Materiel Division, he further inquired as to whether the group was designed to serve multi-engine or single-engine units when it arrived overseas. But, so far as is known, he received no answer to the latter question, and was told that "no determination as to type of Air Depot Group, tropical, temperate, or cold climate, had been made." [17] He was informed that he would be notified as soon as such a decision had been reached.

## COMPOSITION

Originally an Air Depot Group consisted of just three Air Corps squadrons: a Hq and Hq Squadron, a Supply Squadron, and a Repair Squadron. There were no units from other arms and services. Although the groups had been activated to test allotments of personnel and equipment, approved T/O's and T/E's were not released by higher authorities for several months. Instead, approximate strength, composition, and mission were set forth in a letter, dated 18 April 1941, from the Chief of the Field Service Section to the Commanding Officers of Fairfield, San Antonio, and Sacramento Air Depots.[18] The provisions of this letter were by no means definitive. In some operations it was contemplated that Air Depot Groups would be reinforced by additional Supply and Repair Squadrons. In any case, regardless of the number of units to be serviced or the amount of work to be processed, they were to furnish "all phases of supply and maintenance" within their sphere. Total strength for each group was to be 19 officers and 239 enlisted men.

As an indication of the tentative nature of this directive, it should be noted that provision was made for the assignment of other arms and service personnel either as individuals or as units. A list of other arms and service positions was set forth with the explanation that such personnel might be assigned as individuals. If so, they were to be requested for initial assignment in numbers considered to be the "minimum peacetime strength for operations and training." In addition, certain arms and service units were listed tentatively, including: Engineer Companies (Avn), Signal Platoons (Air Base), and three Quartermaster units—Truck Companies, Light Maintenance Platoons, and Supply Companies (Avn).

The Field Service Section, however, had provided neither for the personnel nor for the units. As a matter of fact, other arms and service units were not approved by the War Department for assignment to Air Depot Groups until well after the Louisiana-Carolina Maneuvers, late in 1941, and they were not so assigned until after the groups had been prepared for movement overseas.[19] Trucks, ordinarily provided by the Quartermaster, were furnished by the Transportation Section of Hq Squadron. Signal and Finance matters were handled by Air Corps personnel. Chaplains, although recommended (for many reasons), had not yet been supplied.

The first T/O's for Air Depot Groups were dated 1 July 1941; they applied to the three Air Corps units only. (See Fig. I.) Although the designations of two of the units were changed, the functions of all three remained the same. Authorized strength for the three squadrons together, officers and enlisted men, was approximately 500. Thus, the first group of this type to be shipped overseas, the 4th Air Depot Group,

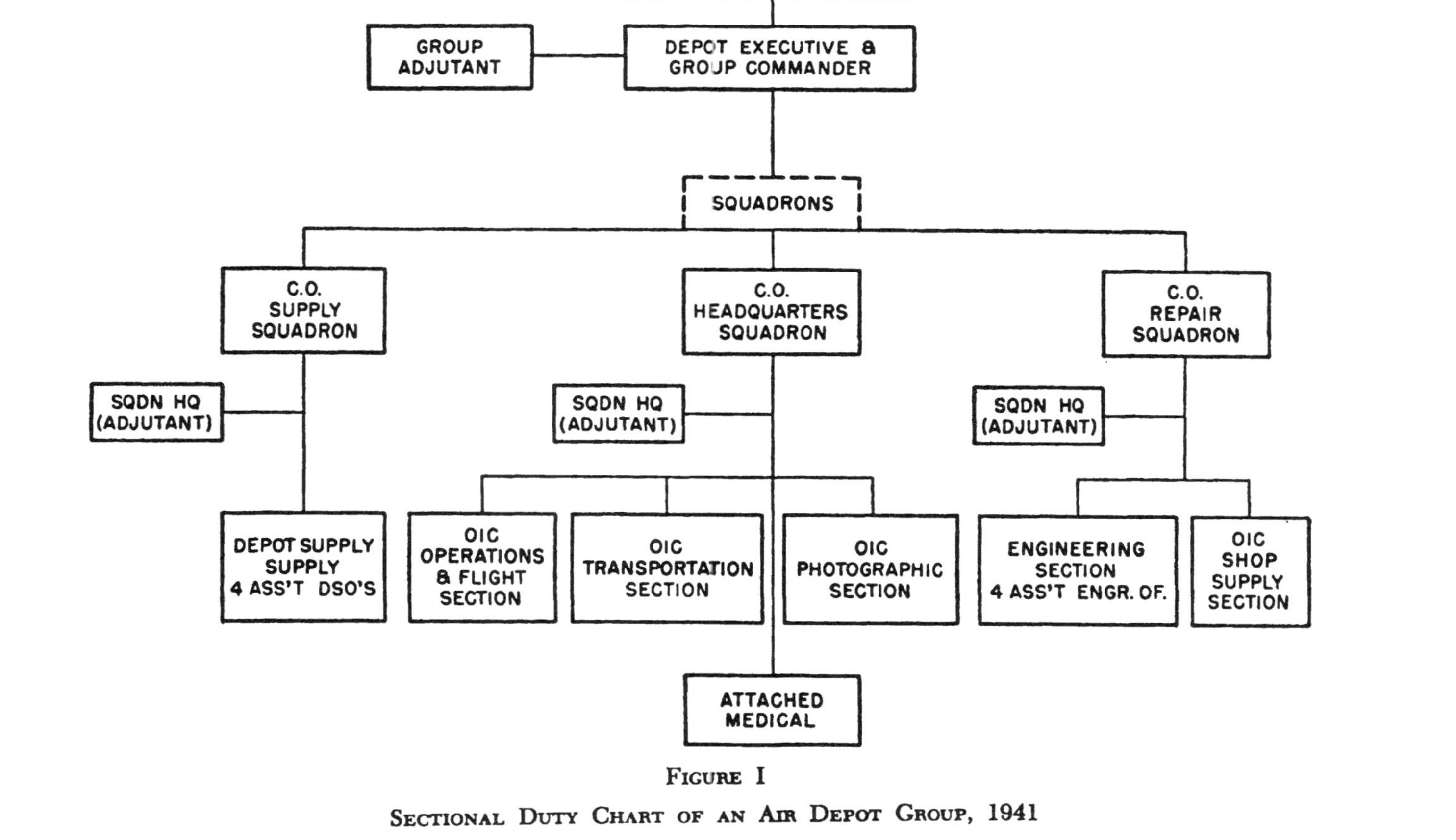


FIGURE I

SECTIONAL DUTY CHART OF AN AIR DEPOT GROUP, 1941

departed from the San Francisco Port of Embarkation on 12 January 1942 with 33 officers and 467 enlisted men.[20]

### Sources of Enlisted Personnel

Pending decision by the War Department on the question of including other arms and services, the strength of Air Depot Groups remained in doubt. Initially 19 officers and 239 enlisted men were authorized by the Chief of Field Services; if personnel of other arms and services were to be included, an additional 12 officers and 186 enlisted men would be assigned to bring the group up to "the proposed peacetime strength of attached services." Not until the T/O's of 1 July 1941 were published was the question authoritatively settled.

During the period of doubt as to the ultimate strength of the groups, requisitions for personnel were ineffective, and training was delayed. Construction projects were also held up until the Engineers could learn how many soldiers there would be per group. Vigorous efforts were made by the Field Service Section to obtain approval of construction for the larger groups, but the Engineers were informed by the War Department that "any strength above the present authorized 19 officers and 239 enlisted men comes out of the Second Aviation Objective, which does not become effective until July 1st." [21] The Second Aviation Objective was an increase in the Expansion Program from 54 to 84 combat groups, not formally announced by the War Department until 23 October 1941.[22]

The original plan for obtaining Air Corps enlisted personnel for these four groups was to start with personnel available at the stations of activation, and to requisition more, when needed, from recruitment centers. Training of recruits would be accomplished first by bringing them together at Jefferson Barracks, Missouri, for screening,[23] and then by sending those who were qualified to Air Corps Technical Schools (later Technical Training Command Schools) at Chanute Field, Illinois, Lowry Field, Colorado, Scott Field, Illinois, and Kansas City, Missouri. Upon completion of courses at these schools, personnel would be assigned or reassigned, as the case might be, to their proper organizations. This plan "presupposed an interval of at least six months prior to the Air Depot Groups being ready for field service." [24] Meanwhile, a small cadre of experienced Old Army enlisted men would be assigned to each of the groups from local Transport Squadrons and Materiel Squadrons to begin the required administration. That this plan was followed in the main can be seen from the fact that the strength of the 5th Air Depot Group at San Antonio, as of 10 October 1941, was as indicated in Table 2.[25]

TABLE 2

STRENGTH OF THE 5TH AIR DEPOT GROUP, OCTOBER 1941

| | | |
|---|---|---|
| Officers | | 7 |
| Enlisted Men | | |
| Present for duty | 138 | |
| Detached Service: | | |
| Mobile, Ala. | 78 | |
| Jefferson Barracks, Mo. | 1 | |
| Chanute Field, Lowry Field, Scott Field, and Kansas City, Mo. | 205 | |
| Normoyle, QM Depot, San Antonio, Tex. | 1 | |
| | | 423 |
| | | 430 |

On 23 April 1941, Colonel Henry J. F. Miller, Chief of the Provisional Air Corps Maintenance Command, recommended that this plan be altered "in view of the contemplated dispatching of task forces outside the continental limits of the United States at an early date." [26] Colonel Miller suggested as an alternative that 140 enlisted men of classifications vital to the maintenance and supply of aircraft be made available for transfer to the four Air Depot Groups from the next graduating class of the Air Corps Technical Schools. Such personnel would be replaced by the recruits who had been assembled at Jefferson Barracks and assigned to Chanute Field for training. The objection to his proposal, no doubt, would be that many of the enlisted men then graduating were undergoing training preparatory to assignment to GHQ Air Force units. But he believed his suggestion was in the best interests of the service as a whole, since Air Depot Groups would be required to assist such GHQ units in task forces overseas. This plan, however, was never carried out.

On 28 April 1941, Colonel Miller made another recommendation to the Chief of the Materiel Division.[27] At that time, he stated, the Maintenance Command was engaged in establishing sub-depots at various stations in the Southeast and Gulf Coast Training Centers, which were autonomous commands over which the Maintenance Command had no control. These sub-depots had been authorized in January 1941: [28] they were being manned insofar as possible with civilian personnel, thereby releasing enlisted men formerly assigned to base engineering and supply departments to other activities under their former commands. He proposed, therefore, that the Commanding Generals of these Training Centers be requested to submit to the Maintenance Command lists of those available for transfer, complete with their qualifications. Personnel thus obtained from the Southeast Training Center would be assigned to the 4th Air Depot Group at Patterson Field; personnel from the Gulf Coast Training Center would be assigned to the 3rd and 5th Air Depot Groups at Duncan Field.

Enlisted men formerly assigned to sub-depots were regarded as specialists because many had had years of training and experience in the journeyman trades. Machinists, sheet metal workers, aircraft engine mechanics, instrument repair mechanics, shop superintendents, and foremen were badly needed. It was Colonel Miller's contention that their employment by the Training Centers in any capacity other than that for which they had had such extensive training would be a waste of valuable manpower.

On 26 June 1941, Colonel Miller wrote to the Materiel Division again.[29] In this letter he stated that his office had received no information as to what action had been taken in response to his recommendation of 28 April. In the meantime, he stated, it had become more than ever apparent, "due to events occurring daily in the international situation," that mobile Air Depot Groups should be brought to a point of efficiency at the earliest practicable date. The groups were beginning to be manned primarily with recruits, contrary to the original plan; what was needed was a nucleus of properly trained enlisted men.

In support of this new request, Colonel Miller attached a list of individuals who had been relieved of their special duties at Barksdale Field, Louisiana, by civilian personnel at the recently established subdepot.[30] These men, he stated, were generally in the lower grades and ratings; if transferred to organizations at Barksdale Field, they would undoubtedly find their places at the bottom of the promotion ladder. If, on the other hand, they could be transferred to the new Air Depot Groups, they would soon be given promotions.

Colonel Miller also stated that the Commanding Officer of Randolph

Field, Texas, was willing to transfer suitable enlisted men to the two Air Depot Groups at Duncan Field. Furthermore, the Commanding General, Southeast Training Center, had approved transfers of several individuals to Patterson Field for the use of the 4th Air Depot Group. Such actions, he felt, clearly indicated that the field commanders were in agreement with his proposal. Accordingly, he resubmitted his request that Training Centers be required to prepare lists of personnel available for transfer to the Maintenance Command.

So far as is known, however, this proposal was not carried out until 26 August 1941, when sub-depots were established at all stations under the direct control of the Chief of Air Corps.[31] At that time it was directed that "enlisted men assigned to the sub-depot, when it is established, will be released as soon as adequate civilian personnel is employed. Enlisted personnel so released will be reported by name and by qualification to the Chief of Air Corps, through the Commanding General, Air Corps Maintenance Command, with a view to their assignment to activities of the Air Corps where their trained services will be utilized."

Another potential source of enlisted personnel was the group of men recruited from civilian occupations at air depots. At Patterson Field, Ohio, the Maintenance Command was particularly fortunate in that the Fifth Corps Area was willing to grant the depot commander authority to do his own recruiting.[32] A vigorous effort was made, with the result that the 4th Air Depot Group obtained at the start a large number of exceptionally well-qualified civilian workers. Later in the year San Antonio obtained a similar authorization from the Eighth Corps Area. It is not known how many were assigned to Air Depot Groups in this manner, but the total was probably less than 500 in the period before Pearl Harbor. Air Depot Groups were authorized between 400 and 500 per group, and there were just four groups in the country, none of which was brought to strength until after Pearl Harbor.

Finally, there were the selectees, inducted into the army supposedly for one year's training: only a few in the early life of these organizations, but more as time went on. In requesting selectees, Colonel Miller defined the qualifications he desired in a list of eleven specialties. Classification numbers were not designated; jobs were indicated merely by one word titles.[33]

This request was approved shortly afterwards. Moreover, The Adjutant General listed detailed occupational requirements, based on the 1942 Troop Basis, which could be requisitioned by Maintenance Command organizations.[34] Specialty Serial Numbers were later set forth, based on paragraphs 4 and 5, Army Regulation No. 615-26, with the number to be requisitioned per group, as shown in Table 3.[35]

TABLE 3

QUALIFICATIONS OF AIR DEPOT GROUP PERSONNEL

| *Specialty Serial Number* | *Designation, Occupational Specialist* | *Number per Unit* |
|---|---|---|
| 006 | Aircraft Mechanics | 10 |
| 008 | Aircraft Engine Mechanics | 4 |
| 050 | Cabinet Makers | 4 |
| 055 | Clerks | 12 |
| 071 | Draftsmen | 2 |
| 098 | Watch or Instrument Repairmen | 2 |
| 114 | Machinists | 3 |
| 121 | Mechanics, General | 4 |
| 144 | Painters | 3 |
| 201 | Sheet Metal Workers | 3 |
| 213 | Stenographers | 10 |
| 215 | Electricians | 2 |
| | | 59 |

Selectees for all branches of the army were obtained from Reception Centers under the nine Corps Areas. For Air Depot Groups, Corps Areas were allotted "procurement to begin on or about June 1 for specified Air Corps units and stations." Requisitions for personnel were to be submitted on WD AGO Form No. 210 to the appropriate Corps Area through the Chief of the Maintenance Command one month in advance of the dates on which adequate housing, hospitalization, and supply facilities would become available.[36] The number of trainees assignable to Air Corps units was exact: each of the four Air Depot Groups was authorized 16 selectees for Hq and Hq Squadron, 25 for the Repair Squadron, and 18 for the Supply Squadron—totaling 59. No colored personnel were mentioned in this allotment.

Thus, enlisted men were obtained from four sources. First, cadres of Old Army enlisted men were assigned from Transport Squadrons and Materiel Squadrons at the bases of activation. Second, enlisted men were obtained, upon replacement by civilians, from newly established sub-depots, originally at stations under the Southeast and Gulf Coast Training Centers, and later at all stations under the direct control of the Chief of Air Corps. Third, additional personnel were recruited from among civilian employees at air depots. Finally, the selectees. Of the four categories only the last was composed of unskilled basics, requiring individual technical training. Men assigned from Transport Squadrons, Materiel Squadrons, and sub-depots already had an extensive experience in the maintenance and supply of military aircraft. Civilians recruited in air depots were likewise picked because of their technical qualifications. Group training for these men consisted merely of amplifying their experience by on-the-job instruction in depot shops. Unfortunately, much difficulty was experienced in obtaining sufficient numbers of any of the above classifications.

### Sources of Officer Personnel

Qualified officers were even harder to locate. For many months it was customary for officers assigned to Air Depot Groups to occupy such positions in addition to other duties at the depot proper. Moreover, at San Antonio on 1 May 1941 the same four officers who were attached to the 3rd Air Depot Group were also attached to the corresponding units of the 5th Air Depot Group.[37] Originally, these were depot officers who had been detailed to the groups as an additional duty. The first Commanding Officer of the 4th Air Depot Group, for example, was Major R. W. Stewart, Adjutant of Patterson Field, who retained both positions for some time. The Commanding Officer of the 2nd Air Depot Group was Lieutenant Colonel A. G. Liggett, Supply Officer at Sacramento Air Depot. He, too, retained both positions.

Throughout 1941 the number of officers assigned to Air Depot Groups remained small. The 4th Air Depot Group went to the Port of Embarkation with but 22 officers; it sailed with only 33.[38] Air Corps Reserve and National Guard officers were scarce, and it was necessary to transfer commissioned personnel from other branches. Thus, a Captain and a Major, both Infantry officers, were assigned temporarily to this group while it was at Macon, Georgia.[39] Officers from the Cavalry, Field Artillery, and Medical Corps were later assigned to this unit at the same location.[40] A few technical officers were obtained by transfer from the Transport Command.

At first, however, the Maintenance Command relied upon Regular

Army, National Guard Air Corps, and Reserve officers, available without recourse to any other command or branch. On 26 June 1941, depot commanders were urged to "give every consideration to the assignment of qualified reserve officers from those now under procurement."[41] Later, an increasing proportion of OTS officers (direct commission men) and OCS officers (former enlisted men) were obtained from Miami Beach. Ultimately, these men became the most important category from which Air Depot Group leaders were picked.

### Group Functions

The mission of Air Depot Groups was first defined on 18 April 1941. The Chief of Field Services on that date instructed Air Depot Group Commanders that their units would be trained and equipped as mobile supply and repair organizations for the support of combat units in the theaters of operations.[42] In the field, Air Depot Groups would serve as advance or intermediate organizations where combat units were far from the established base depots in Panama, Hawaii, and the Philippines. At first they would perform minor phases of third echelon maintenance, and such second echelon maintenance as might be required.[43] Fourth echelon work, such as complete airplane or engine overhaul, would be undertaken only in cases of emergency; normally such work would be shipped to the base depots mentioned above, or to one of the four air depots in the United States. Salvage and reclamation of wrecked aircraft, however, would be undertaken if necessary. One of the most important functions of these Advance Air Depots would be to act as intermediate links in the chain of supply from base depots to combat units.

The new organization was thus to supplement all existing air force service units. Combat groups of the GHQ Air Force in 1940 and 1941 performed only first echelon maintenance, which could be accomplished with kits furnished to crew chiefs, radio repairmen, and armorers. The Materiel Squadrons of Air Base Groups of the GHQ Air Force performed only second echelon maintenance, which could be accomplished with T/BA equipment and the supplies set up in the 72-hour kits of each tactical unit.[44] By performing some third echelon work in the field, therefore, the new Air Depot Groups were accomplishing an entirely new service for the air forces.

Originally the word "mobile" was to have an important bearing on the operation of the group. The group's purpose was to accompany a rapidly moving front, carrying supply shops, technical equipment, and repair tools. It was meant to be entirely mobile, settling down long enough to perform a job, then pulling up stakes to accompany the advancing or retreating air forces.[45] Much emphasis was placed in all early directives on mobility, but when the units arrived overseas, many of them found, as did the 4th Air Depot Group, "the nature of the war, and conditions of operations encountered, rendered it impractical to operate as a mobile group, as was intended." [46]

## The Problem of Engine Overhaul

The functions of Air Depot Groups did not at first include engine overhaul. In September 1941, however, the Chief of the Maintenance Command was directed to take the necessary steps to equip six groups, although the groups themselves had not yet been activated, for the overhaul of engines and accessories at the rate of 100 complete overhauls per month.[47] To comply with this directive it was necessary to initiate procurement for additional buildings, shop equipment, and tools —in addition to requisitioning the personnel.

At first, higher authority labored under the false impression that complete engine overhaul could be accomplished by personnel of the Repair Squadron, in addition to their other work.[48] The Maintenance Command suggested immediately that if this were to be done, the Repair Squadron should be reinforced on the basis of at least three additional men for each engine overhauled per month.[49] The alternative of strengthening the group by one extra Repair Squadron was rejected as impractical. To perform complete engine overhaul would require specialized training. It would be necessary either to provide extra training for the personnel of the Repair Squadrons or to constitute, activate, and train entirely separate Engine Overhaul Squadrons. Yet not until 1 July 1943 was a separate Engine Overhaul Squadron authorized by the War Department.[50] Until then such work was performed by a special Engine Overhaul Section in each Repair Squadron needing it. The simpler procedure of augmenting a group by one extra Repair Squadron, without specialists trained in engine overhaul, was obviously inadequate. Accordingly, that plan was dropped, and the Maintenance Command (later the Air Service Command) undertook to expand the existing Repair Squadrons by furnishing overhaul experts.

On 8 December 1941 the Assistant Chief of the Air Service Command requested that each Repair Squadron be augmented by 125 enlisted men with specified qualifications for engine overhaul.[51] This request was granted, but procurement of the accompanying equipment was not initiated until after the first of the year, and then only for "one half the total of Air Depot Groups to be activated." [52] Priority for Air Depot Groups was established as follows: [53]

First priority—4th Air Depot Group, Patterson Field
Second priority—1st Air Depot Group, Canal Zone *
Third priority—5th Air Depot Group, Duncan Field
Fourth priority—2nd Air Depot Group, McClellan Field
Fifth priority—3rd Air Depot Group, Duncan Field
Thereafter as the groups were activated.

### The Louisiana-Carolina Maneuvers

In 1941 Air Depot Groups had had little opportunity for unit training. Such an opportunity presented itself in the projected maneuvers that were to take place in Louisiana and Carolina from September to November. Unfortunately, the original plans did not include Air Depot Groups. Indeed, no provision was made in the preliminary outlines for any air force service units other than the Air Base Groups of the Combat Command, which at that time did only first and second echelon work. Apparently, third and fourth echelon work was to be done at the continental air depots. The authorities in Washington seemed to assume that these air depots would always be available, and that it would not be necessary in the maneuvers to test tactical Air Depot Groups.

To remedy this deficiency, Brigadier General Henry J. F. Miller and Lieutenant Colonel E. E. Adler of the Maintenance Command made a trip to Washington where they "got down on their knees and begged" for permission to send Air Depot Groups to the maneuvers.[54] They urged that these organizations be given a trial, for they doubted that the Materiel Squadrons and other sections of Air Base Groups, and detachments of service personnel from air bases, would be sufficient for the service requirements of a combat task force at some distance from fixed points of supply. The performance of third echelon maintenance and supply by Air Depot Groups would be the only connecting link be-

* The priority of the 1st Air Depot Group was to obtain personnel and equipment; there was no intention of shipping this group from Panama to any other part of the world.

tween the zone of the interior and the task forces which might later be sent outside the continental limits of the United States. There was no other active unit in the Air Corps which could perform this function.

In response, the Chief of Air Corps authorized participation in the maneuvers by the 4th Air Depot Group and a few units of the 3rd. Thereupon Lieutenant Colonel E. E. Adler, Chief of the Plans Division of the Maintenance Command, published on 14 August 1941 a *Memorandum to All Concerned,* outlining instructions to his units.[55] Although time was short, a great effort was made to enable these units to compare favorably with the facilities offered by the Task Force Service Commands of the Air Force Combat Command, for the Maintenance Command believed that the maneuvers would reveal the deficiencies of the existing system and the necessity for more highly specialized tactical service units.

For purposes of planning and observation the maneuvers were to include four phases. The first phase would be completed in the Louisiana area. The second phase was a move to the Carolina area. The third phase consisted of certain Interceptor Command exercises, and the fourth and final phase, a maneuver in conjunction with large ground forces in the Carolina area.

### THE LOUISIANA PHASE

General instructions covering the first phase provided for two Task Forces designated as the Second Task Force, operating in the Northern Area, and the Third Task Force in the Southern Area. Each of these was provided with a Task Force Service Command whose responsibility included not only Air Corps supply, maintenance, salvage, and reclamation, but also supervision over all other supplies for the units engaged. In brief, the areas were so organized that the entire Service of Supply was under the command and control of the respective Task Force Commanders, who were agents of the Air Force Combat Command. This arrangement was appropriate, as it was desired to give a thorough test to the supply system of the Combat Command.

Each Service Command included an air base. One was located at Jackson, Mississippi; the other at New Orleans, Louisiana. Scattered laterally through the area assigned to the two air bases were sub-bases, with distribution points at all of the operating airdromes. The Second Task Force Service Command Headquarters was located at Natchitoches, Louisiana, and the Third Task Force Service Command Headquarters was at Lake Charles, Louisiana. The 4th Air Depot Group was to be in the curious position of serving as an Advance Air Depot for both Service Commands.[56]

More specific instructions directed the 4th Air Depot Group to move from Patterson Field, Ohio, to Jackson, Mississippi, so as to arrive on 26 August 1941. The group was to receive housing and fixed shop facilities from the Commanding General, Third Air Force, who was in charge of the entire maneuver area. It was instructed to establish an initial thirty-day level of supplies for both Service Commands, and to maintain that level if possible throughout the exercises. Requisitions were to be submitted for this purpose directly to the Fairfield Air Depot in Ohio. Supplies would be distributed only to the Jackson and New Orleans air bases, on requisition from the respective Service Command Headquarters of the two Task Forces.

It was determined as a principle of supply for this particular exercise that it would be necessary to issue complete new airplanes to replace those grounded for lack of spare parts. The parts were not available, neither could the time be found for third and fourth echelon maintenance. The 4th Air Depot Group was to operate an Air Park as a receiving point for new airplanes from the "zone of the interior." Task Force Service Commanders were to be notified of the number of replacement planes arriving, while crews to fly the planes to their battle stations were to be provided by the Combat Command.

First and second echelon maintenance, as defined in existing Technical Orders, would be performed by Materiel Squadrons of Air Base Groups under the supervision of the Service Commands of the Task Forces. The function of the Air Depot Group was to accomplish such third echelon work as might be referred to it by the Service Commanders. It was decided, however, that third echelon repairs in the field would have to be restricted. Small airplanes, not easily repairable in the 4th Air Depot Group, would be shipped to a control depot, as designated by the Field Service Section of the Maintenance Command. Large airplanes such as B-18's, B-25's or B-26's would be repaired where they fell and flown to an air depot.

A blanket invitation was extended to manufacturers[57] to send their representatives to the maneuver areas to assist in disposing of maintenance difficulties pertaining to their particular types of equipment. They were directed to report initially to the Commanding Officer of the 4th Air Depot Group to receive authorization to visit such parts of the area as they desired. Any inconveniences encountered by them in the course of their visits would be reported in writing to the Commanding Officer of the 4th Air Depot Group.

Miscellaneous instructions for the maneuvers included directives that funds required for the operation of the Advanced Air Depot would be requisitioned through the Maintenance Command, and the 50th Transport Wing would attach three transport airplanes complete with crews to the 4th Air Depot Group. The 50th Transport Wing was also requested to furnish such additional emergency transport service as was called for by the 4th Air Depot Group.

Questions about the chain of command remained unsolved until a few days before the maneuvers actually began. Major General George H. Brett, Chief of the Air Corps, devoted a great deal of time to coordinating the ideas of the several commands on this subject. On 15 August 1941 he wrote two memoranda, one to the Chief of the Maintenance Command, and one to the Commanding General of the Air Force Combat Command, "confirming conversations of recent date and the conferences of August 11" (which had been attended by the three general officers and members of their staffs). In these memoranda he set forth his understanding of the conclusions which had been reached as to the supply and maintenance phases of the maneuvers and the "support to be rendered to the Air Force Combat Command by activities operating under my control." In one he said:

> The Fourth Air Depot Group . . . will be considered to be an Advance Air Depot extending the normal services of the control depots of the zone of the interior to close proximity with the combat forces. As an Advance Depot, it will remain under the control of the Chief of Air Corps operating as a zone of the interior establishment under the control of the Chief of Service. Accordingly, it will not come under the command or control of the Combat Command. It will support these activities.[58]

There was little time to prepare the 4th for its move from Patterson Field. Orders were not published until 14 August, and the group had to leave on the 21st. The result of this haste was considerable confusion and a rather poorly organized 87-vehicle motor convoy. On the first night out personnel were not bedded down until well after midnight.[59] Not until the third day were automotive repair crews placed at the end of each serial of the convoy, equipped with the tools and supplies to keep the vehicles on the road. As soon as this was done, however, stragglers were repaired and pushed into camp at the end of each day without too much delay. It was found that 200 miles per day was all that could be covered by a convoy of this sort, even with an ample stock of repair parts.[60]

The principle problem of the group, once it was located at the Jackson (Mississippi) Airport, was the matter of supply. Fifty freight-car loads of equipment and supplies had been shipped to Jackson for the use of the group, and the size of this shipment had created the impression that every tool and spare part imaginable had been included. No one realized, apparently, how quickly shortages could develop. No arrangements had been made for a daily air transport service from Patterson Field; the assigned airplanes from the 50th Transport Wing were used locally. And yet, by the end of the first week of maneuvers, only 8–10 percent of the airplane spare parts being requisitioned on the 4th Air Depot Group could be supplied. Another surprising situation arose from the fact that of all the requisitions received, approximately 75 percent were for items not needed in the maintenance or repair of airplanes —pencils, paper clips, and toilet paper. Moreover, large quantities of small parts common to all aircraft had to be obtained because many of the units in the area were poorly organized and did not have complete sets of their own organizational equipment.

To house its vast assortment of miscellaneous items the group had only one hangar, which it shared with Lieutenant Colonel Paul E. Ruestow's 3rd Materiel Squadron. Unloading, storing, recording, and arranging of supplies for issue presented many new problems. For example, a shipment of P-39 parts came directly from the factory, through the Fairfield Air Depot, without being unpacked or marked or tagged. Even with the expert help of Mr. Alt of the Bell Aircraft Company, this situation caused a delay of several days, for personnel of the group were slow at tagging the parts.*

Perhaps the chief weakness of the 4th Air Depot Group, however, was

* For further insight into the supply problem of the 4th Air Depot Group, note the list of property compiled from the group's stock record cards as of 5 Oct. 1941, showing balance remaining after the Louisiana Maneuvers and the amount used in the maneuvers, in TSAGD, Microfilm Unit, 1941 Correspondence, Reel 14, Item 25. *This was the first information the AAF had, based on practical experience, which could be used for estimates of unit requirements.*

the inexperience of key personnel. Except for a Major, a Captain, and a Warrant Officer, every officer in the group was a Second Lieutenant.[61] Only one, Lieutenant Wells, who had commanded the group before it went into the field, had had previous experience in supply. The report of the commander of the group stated that "the organization could not have functioned outside the continental limits of the United States with the personnel originally assigned." All reports on the maneuvers emphasized the need for more careful consideration of the technical nature of the tasks involved, and the selection of officers whose qualifications and previous experience justified their assignments.

Lack of previous military experience was particularly hampering in Air Corps supply. Engineering officers of bases, sub-bases, and distribution points were unfamiliar with existing supply procedures, and unable to obtain the equipment they needed. Other difficulties listed by one of the inspectors were:

Lack of coordination between the Air Force and the Maintenance Command on compiling tables of allowances for spares to be carried by tactical organizations in the field.
Incompleteness of the 72-hour and the 10-day kits as made up by the Air Force Combat Command.
Lack of stock catalogues amongst the various activities on all types of new aircraft.[62]

The Service Commands of the Air Force Combat Command were also severely criticized. Requisitions were received by the 4th Air Depot Group from the Service Command installations at Jackson and New Orleans, and supplies were delivered to these two air bases by air and by truck. This system, unfortunately, required redelivery and redistribution to sub-bases and distribution points. If the handling organizations had been reduced in number, and if supplies had been distributed directly, many delays might have been avoided. The comments of Lieutenant Colonel Barney M. Giles, of the Inspection Division of the Office of the Chief of Air Corps, were indicative of the conclusions reached by the Air Corps on the organization established by the Combat Command.

The supply system as established for the maneuvers was considered to be unsatisfactory, due to the length of time required for a requisition to pass through the intermediate agencies in the chain of supply. Namely, a squadron requisition would go to a Distribution Point, thence to a sub-base, thence to a main base, thence to an Air Depot Group, and in most cases the majority of the items requisitioned were located at the rear echelon. Organizations were not notified of action taken on their requisitions, and whether or not supplies requisitioned would or would not be supplied.

During these maneuvers the terms "10-day supply" and "30-day supply" were used as common terms and the sub-bases were notified that they would be stocked with 10-days supply. A number of visiting officers and umpires heard the term "10-day supply" and "30-day supply" used, and it was generally understood that such supplies were available within the maneuver area. As a matter of fact, the term "10-day supply" and "30-day supply" for the Air Corps is very confusing, since there are no lists made up indicating what constitutes a 10-day supply or a 30-day supply for any of the modern type airplanes. This resulted in considerable confusion, since a number of airplanes were actually out of commission awaiting spare parts.[63]

It was also reported that the communications system between the two Service Commands and the 4th Air Depot Group operated poorly. Indeed, "the Signal Officer of the Second Air Task Force Service Command apparently had the telephone company in such confusion," according to Major Warren, "that they did not even know the proper organizations or locations for installation of TWX machines." No directory was ever published as to the correct name or number of the several TWX machines. At Jackson there were four such machines and "it was seldom that a message addressed to any one of the four organizations was received on the proper machine." In addition, TWX machines were restricted to emergency requests for urgently needed supplies, as a result of the lack of funds, so commercial telegraph and army radio facilities had to be used also. On commercial telegraphs every message was subjected to a delay of from two to ten hours. On the army radio net, in one case, three days were required for the delivery of a single message.

Maintenance difficulties were not so pronounced as those of supply during the Louisiana Maneuvers, partly because of the limited nature of the repair work undertaken. Thirty-two wrecks were reported to the 4th Air Depot Group; all, with the exception of one P-39, were shipped to the San Antonio Air Depot. Work which had to be accomplished at the crash site was accomplished by civilian technicians from the control depots.[64] Although engine change and other second echelon functions were performed by the 4th, no modern engine-mount assemblies were used, and in several cases airplanes were kept out of commission indefinitely awaiting engine change. Under the circumstances, the general standard of maintenance aircraft was considered "very satisfactory" by Colonel Giles.*

As a result of the Louisiana maneuvers, the group engineering department became the subject of several important recommendations. One involved the transfer of the Flight Test Section from Hq and Hq Squadron to the Repair Squadron. This recommendation had been made while the maneuvers were still going on by Lieutenant Colonel Clements McMullen, Commanding Officer of the San Antonio Air Depot, who had been in a good position to observe the group's work, and it was repeated later several times by him and by others,† but was not finally put into effect until July 1943.[65]

Colonel McMullen also recommended that Field Services should "double the size of the present Repair Squadron," because "at its present strength [it] is scarcely able to carry on second echelon maintenance for more than four groups." The criticism implied in this remark was that third echelon work was rarely attempted. Major Warren was less severe in his discussion; and for the condition referred to he offered the following explanation:

> It is almost impossible to make any recommendations in regards to the increase in number of enlisted men within the three organizations [of an Air Depot Group], as the information gathered from the Jackson Maneuvers does not present a true picture of the use of the Air Depot Group. It is my understanding that an Air Depot Group was designed to handle two or three combat groups, and not eight as was true in the case of the Fourth Air Depot Group during these Maneuvers.[66]

Colonel McMullen recommended further that the Supply Section be removed from the Repair Squadron and put in the Supply Squadron.

* Very Satisfactory is a technical term in army language which really means "pretty poor" or "quite disappointing." It comes between Superior and Excellent, on the one hand, and Unsatisfactory, on the other.

† Lt. Col. Clements McMullen to Chief, Maintenance Command, "Changes in T/O for Mobile Air Depot Groups," 22 Sept 1941, TSHIS-2 files. On 24 Sept 1942, Lt. C. B. Sweeney, Acting Assistant Adjutant General, Hq ASC, wrote to Brig. Gen. Clements McMullen: "This Headquarters concurs in the idea that the Flight Test Section now set within the Headquarters Squadron should be part of the Repair Squadron. After the maneuvers last year in the South, Colonel Warren recommended that the Flight Test Section be transferred from the Headquarters Squadron to the Repair Squadron, and tables were made up accordingly. However, Washington did not concur in this recommendation, and the Flight Test Section was put back into the Headquarters Section." In TSAGD 322 (Air Depot Gps-General).

It had been operating as a Local Issue Section in much the same manner as the corresponding section of an air depot. The proper location for this particular activity, as it happens, both in the mobile group and in the stationary air depot, has often been a matter of hot debate. The transfer of local issue to the Supply Squadron was another change that was not finally adopted until 1 July 1943.[67]

Several other comments, in addition to those above, were made by Colonel Giles. These included criticisms of the maneuvers for not including the firing of guns, the dropping of bombs, or the testing of oxygen facilities for high altitude flights. Changes were recommended in the organization of combat squadrons and groups, and alterations proposed for the general disposition of air force units. Suggestions wholly or partly applicable to Air Depot Groups were the following:

1. That one system of traffic control be used at each field, regardless of activities operating therefrom, civilian as well as military.
2. That the T/O for motor transportation be revised to authorize at least five additional quarter-ton trucks (Jeeps) for each squadron.
3. That consideration be given to procurement of a light tractor tug for use in the field instead of the wheel tug which was satisfactory on hard surfaces only.
4. That certain mobile equipment be provided: namely, armament maintenance trailers, technical supply trailers, and trailers designed to accomodate field kitchens.[68]

Another important recommendation at the conclusion of the Louisiana maneuvers was that Air Depot Groups include detachments of other branches of the service, especially Weather, Quartermaster, Signal, Ordnance, and Engineer Sections. On this proposal there was general agreement.

### THE CAROLINA PHASE

Immediately after the Louisiana phase, the participating air force units moved to the East Coast for continuation of the exercises. The move itself was the second part of the larger maneuvers. In the East, the schedule included both air force exercises and maneuvers in conjunction with the ground forces, including the following: [69]

| | |
|---|---|
| 1. First Army Maneuvers: | 3 October–15 November |
| 2. First Interceptor Command Exercise: | 9–16 October |
| 3. Third Interceptor Command Exercise: | 20–26 October |
| 4. First Army—Fourth Corps Maneuver: | 15–30 November |

Observation units taking part in the First Army maneuvers were scheduled to arrive at their new stations on or about the third of October 1941. They were to occupy the same fields which would be used in the First Army—Fourth Corps maneuvers. To provide continuity, the administrative organization of the Carolina area was started in the middle of September. Most services were provided by the Third Air Force, although the First Army was responsible for furnishing Class I and III supplies (subsistence and fuels) through regular Ground and Service Force channels. Air service elements from Louisiana were moved to the Carolina area at the direction of the Commanding General, Third Air Force.[70]

The First Army-Fourth Corps maneuvers were the fourth and final phase of the entire exercise. During this phase air force units accompanied some 300,000 ground force troops from the Fort Bragg-Camp Jackson area in the largest war-games that had ever been held in this country. So far as the air forces were concerned, operations were based on the same principles as those which had governed in September.[71] One important difference, however, was that the colder climate in North Carolina and the more advanced season made it necessary to provide more comfortable living conditions for the troops. Consequently, wherever possible, organizations operating from fixed locations were provided with tent stoves and suitable bathing facilities, including hot and cold running water.

In these maneuvers Langley Field, Virginia, was the air base for the first Air Support Command, and Daniel Field, Augusta, Georgia, was the air base for the Third Air Support Command. Sub-bases in the First Air Support Command Service Area included: (1) Cape Lookout, N.C.; (2) Roanoke Rapids, N.C.; (3) Roanoke, Va.; (4) Marion, Va.; (5) Salisbury, N.C.; (6) Chesterfield, S.C.; (7) Georgetown, S.C.; and (8) Wilmington, N.C. Sub-bases in the Third Air Support Command Service Area included: (1) Savannah, Ga.; (2) Georgetown, S.C.; (3) Chesterfield, S.C.; (4) Salisbury, N.C.; (5) Marion, Va.; (6) Asheville, N.C.; (7) West Minster, S.C.; and (8) Milledgeville, Ga.

To service tactical units at these installations, and in the rest of the North Carolina area, the 4th Air Depot Group was moved from Jackson, Mississippi, to Macon, Georgia (Herbert Smart Airport) in the early part of October 1941.[72] Its functions were much the same as before, except that the system of supply was changed. Under the new system the air base at Langley Field maintained a thirty-day level of supplies for units of the First Air Support Command, while supplies for the Third Air Support Command were divided, a ten-day level being maintained at Augusta, and a twenty-day level at the air depot at Macon. These various levels, incidentally, had still not been defined in terms of given numbers of parts for particular airplanes.

Far from the scene of "action," the 4th Air Depot Group quietly concentrated on the tasks assigned to it—not least of which was the transportation of bulky supplies and equipment from Jackson, Mississippi, and the establishment in Macon of an Advance Air Depot. Seventy carloads of freight were dispatched as the train echelon of its initial shipment,[73] and this feat was accomplished in spite of the fact that the group had no Quartermaster personnel and not a single officer who was familiar with the preparation of government bills of lading.[74] In gauging the difficulty of this task, it must be remembered that limitations imposed by financial responsibility were undoubtedly greater in the prewar days than they would have been later, after dollar values had become less important. Instructions issued in a teletype from Headquarters were quite specific:

> . . . Funds in the amount of $14,777.50 were allotted to cover movement to Jackson, Mississippi, and to Macon, Georgia. They were to cover all possible contingencies with the understanding that sufficient savings would accrue to cover return movement to Patterson Field. . . . Authority may be granted to over-obligate specific-purpose numbers: which authority must in all cases act to decrease the allotment in other specific-purpose numbers. However, it will be the responsibility of the Commanding Officer, 4th Air Depot Group, to insure that sufficient savings do accrue to cover the return movement of the 4th Air Depot Group to Patterson Field.[75]

In view of the shortage of personnel for loading and unloading, the planning of such a shipment was not easy.[76] Railroad regulations required heavy demurrage charges if the using agency was unable to load or unload freight cars within forty-eight hours after the company had "spotted" its cars for that purpose. Fortunately, in Jackson the railroad company could give the group only four or five cars a day, and Major Warren was able to procure several extra strong backs from the San Antonio Air Depot. In Macon there were but two sidings with space for only ten cars on one platform, and no platform at all on the other track, so the company was able to "spot" only ten cars at a time for unloading.

This combination of circumstances enabled the group to escape demurrage charges, which would otherwise have been unavoidable.

In addition to the train echelon, there was also a motor convoy. One of the convoy's troubles was the problem of "beating the whole Third Army" through Montgomery, Alabama.[77] Major Warren knew that on his way to Georgia he would have to cross the path of the Third Army on its way to North Carolina, and he knew that it would take the Third Army at least four days to clear Montgomery. But he was unable to get specific information on their schedule because "someone was afraid he would tell the enemy." Great initiative and not a little mind-reading were required to obtain the necessary information. Major Warren's reasoning, as indicated in a transcribed telephone conversation, was as follows:

> . . . we believe they are not any of them bivouacking in the Montgomery area, as we suspect there is a bivouac area for each serial as it moves through. They will probably have the same bivouac areas to keep them separated, see? . . . We have called them at Lake Charles and Colonel Drane has sent an officer up to see Colonel Lutes (who is the G-4) to try to find him, because we can't reach him by telephone to obtain this information for us—and also information for the Third Service Command. . . . If they do bivouac fifty to seventy-five miles southwest of Montgomery—if we stay overnight at Montgomery we can be out of Montgomery at daylight, and be about three hours ahead of them.[78]

Other unanticipated problems arose on this journey. One was the result of a practice, developed on the trip from Fairfield to Jackson, involving the use of reconnaissance parties, or advance details, to precede the convoy and smooth its path. It was the responsibility of these scouts to plan the itinerary of the convoy and to avoid bad roads. They were also to coordinate traffic matters with the local police forces in the towns along the way. The duty of conducting such expeditions was rotated among the officers of the group. One inexperienced 2nd Lieutenant, however, neglected to measure an overpass,[79] so several large trailer trucks got stuck, and a whole serial was delayed for a number of hours.

Once in Macon, the task of the group was to set up a new tactical air depot. Facilities were at a minimum, for the 4th Air Depot Group was the first military organization of any kind to utilize Herbert Smart Airport.[80] Rows of tents were pitched in a cornfield—only to be flattened by a hurricane two days later. The three airplanes from the 50th Transport Wing were not damaged, for they were lashed down and sheltered by a circle of trucks. The following day construction was resumed. The entire job was completed for less than $5,000, which was surprising since the much less elaborate camp site at Jackson had cost the group $3,750.[81]

In Macon, as in Jackson, the group performed few repairs on aircraft. The maintenance personnel were kept busy improving the camp and repairing the trucks damaged in the underpass. Airplanes requiring more than first or second echelon work were shipped to the Middletown Air Depot near Harrisburg, Pennsylvania. Although six or eight wrecked airplanes were reported to the group, third echelon work was attempted on only one (an L-1) while the group was at Herbert Smart Airport. No engine overhaul was performed.

So that the group would receive maximum benefit from the maneuvers, the Materiel Division had directed that it was to be supplied with full Organizational Equipment List (OEL) and Table of Basic Allotments (TBA) equipment. In addition, the commanding officer was directed to submit a list of other recommended items, which would be supplied if possible. Captain J. D. Howe, in command of the group at the end of the maneuvers and destined to take it overseas, was unremitting in his efforts to obtain all available equipment. Largely as a result of his endeavors, and the support he received from higher headquarters, "the material and equipment were 85% complete." [82]

### EFFECT OF WAR ON THE TRAINING OF AIR DEPOT GROUPS

As soon as the maneuvers were over, the 5th Air Depot Group was transferred from San Antonio, Texas, to Macon, Georgia, to take the place of the 4th. There was then a probability that the 4th might be ordered overseas immediately; instead it was directed to remain at Macon until after the 5th arrived.[83] This directive was not complied with, however, for the 4th departed from Macon at six o'clock in the morning of 6 December 1941 and the 5th arrived in the afternoon of the same day.[84] Thus, the 5th never had an opportunity to understudy what might be called the original "parent group."

Not long after its arrival in Macon, the 5th was joined by 200 recruits from Brookley Field, Mobile, Alabama.[85] Meanwhile, the 3rd Air Depot Group at San Antonio was directed by Colonel McMullen to set up its equipment at Stinson Field, Texas, as a practical working model, to provide a training location for future groups and to service nearby fields.[86] The 3rd was to learn its functions while teaching others. The 2nd Air Depot Group at Sacramento was still in the process of receiving personnel.

While these plans were being worked out, news arrived of the Japanese attack on Pearl Harbor. Headquarters in Washington immediately undertook to ascertain the status of all active units in the country and learned to its surprise that many changes had taken place in the field which had not as yet been reported. On the day after Pearl Harbor, Lieutenant Colonel Lyman P. Whitten, Office of the Chief of Air Corps, expressed astonishment in a telephone conversation with Colonel James T. Morris and Major Max H. Warren at Wright Field, Ohio, not only that the 5th Air Depot Group had moved to Macon, but also that the 4th had not yet returned to its proper station at Patterson Field.[87] Colonel Whitten was also surprised to learn that a large percentage of the personnel of the 5th was still away at technical schools, and that neither the 4th nor the 5th was ready for immediate shipment overseas.

The purpose of Colonel Whitten's call had been to obtain information so that General Fairchild and General Spaatz could decide the disposition of the two groups that very day. It was clear that decisions would have to be delayed. Major Warren, who had commanded the 4th Air Depot Group during the first part of the maneuvers, explained that that group would not be able to pack its equipment and be on the road in less than a week, and that the 5th would require a good deal longer. Not only were personnel of the 5th still away at schools, but its equipment, in storage at Mobile, Alabama, lacked many essential items. Colonel Whitten had been under the impression that a certain Project 16, then being packed for shipment at Patterson Field, contained a complete set of extra Air Depot Group tools which could be used to supplement the equipment of the groups ordered overseas. He now learned that Project 16 was short precisely the same critical items that were missing in the 5th, and that engine overhaul equipment would not be ready for any of the groups until after 15 January 1942.

Whitten: I would think from what General Arnold and everybody has been told, up here, that the 4th and 5th, both, would be able to be completely equipped, within a week or two, by stealing back from Project 16. . . .

Warren: Well, now, as I just told you, the stuff that is minus in the 5th is also minus in Project 16. . . .[88]

The upshot was that authorities in Washington decided on 8 December 1941 to send the 4th Air Depot Group overseas as soon as it could be prepared to move, and to keep the 5th at Macon temporarily. The 4th was directed to set up its own property accountability for property then held on memorandum receipt, including Quartermaster, Signal, Chemical Warfare, and Ordnance property, and to obtain physical possession of its organizational equipment which was held by accountable officers at Patterson Field, Ohio.[89] It was believed that the training obtained by the 4th during maneuvers was sufficient to justify sending it to an active theater, but that none of the other groups would be ready until they had had more training and practical experience.

The departure of the 4th Air Depot Group from Patterson Field on 15 December 1941 was an unforgettable occasion for those who were present. Everyone knew that this group had always been the pioneer, that the declaration of war meant that many other groups would soon follow it. Brigadier General Henry J. F. Miller spoke briefly and to the point:

Men, it's good to see you. Remember what I told you at Macon, Georgia? Well, I'm going to do it. I'm sending you overseas. Remember, you are my first, my finest group. And I've done and will do everything to see you through, and I know you will not let me down. I want to hear reports of the 4th Air Depot Group—reports as good as those of the maneuvers in the South. Good-by, men. I wish I could go with you. God bless you.[90]

Instructions issued to the 5th Air Depot Group on 8 December 1941, following Colonel Whitten's telephone conversation with Wright Field, indicated the amazing confusion of war plans at that time:

The Fifth Depot Group now moving to Macon will be held at Macon, brought up to strength as rapidly as possible, and every effort made to secure its shortages in equipment. Its assignment is indefinite, but at present it is tentatively earmarked for the Philippines, but may move either east or west to arrive at that destination. A higher priority assignment may arise for this group at Natal. Should the situation change materially, the equipment of this group or perhaps the group itself might go to the Middle East. In any event, it will be held for the present at Macon, and its preparation for movement and action will be expedited.[91]

On 12 January 1942 the 4th Air Depot Group, with Captain J. D. Howe in command, left the San Francisco Port of Embarkation for a destination in the Far East. At the last minute a large number of green officers and raw recruits were assigned; in many ways the organization was far from the standards later required as a matter of routine by POM inspectors. For about a year the several units of this group served in various capacities at locations scattered across the continent of Australia.[92] In October 1942, under Lieutenant Colonel Victor E. Bertrandias, the group was reunited to establish an Advance Air Depot at Townsville, Queensland, Australia, similar to those it had set up during maneuvers at Jackson and Macon.

The remaining groups of necessity devoted the short time at their disposal to preparations for overseas movement. Neither the 3rd nor the 5th was ready to sail until March 1942. Both received last minute additions of recently authorized personnel. "Colonel Dunton told me," wrote Colonel Lester T. Miller, "that when the 3rd Air Depot Group moved away from San Antonio, the men from the other branches arrived one day prior to the movement of the group, and that they were just a mob of untrained enlisted personnel with no equipment and no officers."[93]

To a large extent all four of these groups were models for the later development of the program. The 4th in particular was used for experimental purposes throughout its history. Later units were not to be so fortunate as to participate in full-scale army exercises, or even to service tactical units; yet they would be expected to be cognizant of the lessons learned at maneuvers, and to be able to perform satisfactorily in the theaters of operations. For these reasons the experiences of the early groups were carefully watched. Although sizable numbers of officers and enlisted men actually joined each of the first four groups at the Ports of Embarkation,[94] they functioned efficiently and in many ways had greater advantages than any of the later groups.

# 3

# THE EIGHTY GROUP PROGRAM
# JANUARY–JULY 1942

## INITIATION OF THE EIGHTY GROUP PROGRAM

THE day after the Japanese attack on Pearl Harbor the Air Service Command started a program of expansion which included among its first objectives an immediate increase in the number of Air Depot Groups. Teletypes were sent to the commanding officers of each of the depots requesting information as to the number of additional military personnel that could be accommodated by them in housing facilities then existing or under construction.[1] It was learned that facilities for housing Air Depot Groups were as follows: [2]

| | |
|---|---|
| Fairfield, Ohio | 1 group |
| San Antonio, Texas | 2 groups |
| Sacramento, California | 1 group |
| Mobile, Alabama | 1 group |
| Macon, Georgia (Herbert Smart Airport) | 1 group |

In spite of the discouraging situation revealed by the survey, eighty Air Depot Groups were activated by the beginning of February 1942. The inauguration of this program was accomplished in three rapid steps. Late in December authority was issued for the activation of the 6th to the 11th Air Depot Groups.[3] Early in January authorization was granted for the activation of the 12th to the 40th,[4] and on the last day of the same month immediate activation was ordered for the 41st to the 80th.[5] By 7 February 1942 these eighty groups were located on paper or in fact at the several air depots, or at stations later designated as Air Depot Training Stations.

The haste with which the activations had been ordered gave rise at the outset to many problems. Shortages of housing, personnel, training facilities, and equipment were but a few, and the effect was increased by the fact that the Air Depot Group program of the Air Service Command was well ahead of the expansion of the Army Air Forces. When the

United States entered the war, the 84 Group Program of the Second Aviation Objective was still the army air arm's official guide for procurement and supply. The 54 groups of the preceding program were then 95 percent equipped.[6] Yet eighty Air Depot Groups were activated in January and February of 1942, and each was charged with the servicing of "a minimum of three air force tactical groups."[7]

In December 1941 the Air Service Command had been unable to obtain an exact statement of the need for Air Depot Groups under the 84 Group Program, with the exception of verbal instructions, informally rendered, by Colonel Aubrey Moore of the Plans Division, Office of the Chief of Air Corps.[8] Moreover, procurement under the so-called Victory Program was considered sufficient for the next two years.[9] This program, initiated as an over-all schedule of procurement, in January 1942, was the result of top-level Allied policy decisions, but in many of its details it utilized the work of Lieutenant Colonel Joseph H. Hicks, Jr., Chief of the Supply Branch of the Air Force Section at Wright Field, Ohio, who had long worked for the establishment of a "far-sighted" program. Under its provisions a two-years supply of organizational equipment of all kinds was to be placed on order by 1 April 1942. When the program was started, however, the number of Army Air Force groups foreseen for a period of two wartime years was only 224,[10] which was considerably less than the number adopted in September 1942 as the immediate objective of the 273 Group Program.[11] Nevertheless, in February 1942, the Air Service Command was anticipating the activation of as many as 147 Air Depot Groups in the same two-year period.[12]

The 147 Group idea was quickly dropped. There have never been 147 Air Depot Groups, or even 100; there were shortages enough to plague the Eighty Group Program, especially the shortage of personnel. Complete groups with full T/O strength throughout training were almost non-existent,[13] and it was hard to plan training schedules, or even to set departure dates, until the personnel became available. This situation was to prove a constant source of embarrassment to the command in its program of military training.

### Acquisition of Facilities

The first groups to be manned were housed in tent camps which were far from satisfactory.[14] Preliminary steps had been taken in 1941 to provide the proper facilities for Air Depot Groups, but by the beginning of

1942 these groups still lacked both living quarters and training buildings. In the newer areas the situation had become, or was threatening to become, critical. On 16 February 1942 the Engineer Section reported that:

Much delay has already been occasioned by the fact that the housing for Depot Groups, which was requested last year has not, to the knowledge of this office at this date, progressed beyond the stages of consideration by higher authority. This condition prevails despite the fact that eighty Air Depot Groups have been activated and the personnel is flowing into our depots at the rate of several thousand per month where they will be trained and either sent to the Theater of Operations or a staging area for further training or functioning as a unit. . . . The responsibility for this entire program lies with the Air Service Command, and with the exception of the actual enlistment of the troops composing its strength, every matter affecting a Depot Group is handled by the Air Service Command. It is believed to be in the best interests of the Air Service Command to include this item in the next available Budget Estimate.[15]

On 23 February 1942 authority was granted to proceed with the construction of housing and allied facilities for twenty-two Air Depot Groups. Funds in the amount of $21,194,000 were authorized from Project 1–27 Public 353, Third Supplemental National Defense Appropriation Act, fiscal year 1942.[16] This plan, however, was formulated when only forty Air Depot Groups were authorized. Additional funds, therefore, in the amount of $87,000,000, were included in Supplemental "E" Estimates for the construction of facilities for fifty-eight more Air Depot Groups.[17] It was stipulated that housing for the original twenty-two groups would be used at Air Depot Training Stations in this country, not only for these groups, but also for Transport Squadrons and in some cases for base personnel.[18]

Plans and specifications were made by the Air Service Command for model layouts, which were later to be used at most of the stations at which Air Depot Groups were located.[19] These drawings indicated the utilization of two portable Butler prefabricated hangars (120 x 160 feet) for maintenance activities, and fourteen portable prefabricated warehouses (40 x 80 feet) for supply activities—the minimum technical building requirements for each of the original twenty-two groups. Portable test blocks were added later. The plans also provided for housing of standard mobilization or cantonment type buildings, or portable, prefabricated CCC type buildings.[20] Demountable housing of the CCC type was generally preferred, especially if made entirely of wood, as wood in all likelihood could be replaced in the field when damaged.[21] It was believed that each group would require a total of 40 to 50 acres of ground, of which from 20 to 30 acres would be used for housing.

Supplemental Estimate "E," fiscal year 1942, allocating $87,000,000 for facilities for fifty-eight Air Depot Groups, specified that none of these buildings and hangars were to be constructed in the United States. In other words, they could not be used in the continental training program; rather, they provided overseas facilities, including housing for 1,300 men, with landing field installations and gasoline storage equipment for each group. Contractors were requested to ship these materials directly to Ports of Embarkation.[22] Inspection of such purchases was made at the plant of the contractor before actual shipment. Deliveries were rapid, and orders began to flow to the ports in June, July, August, and September of 1942. Later the deliveries of prefabricated buildings ran so far ahead of the shipment of Air Depot Groups that it became necessary to store these goods for the time being in domestic warehouses. Stored materials were then shipped automatically to the using organizations as needed.[23]

Thus, housing facilities became items of Air Corps supply, rather than real estate facilities for which accountability was maintained in this country. As portable warehouses and prefabricated hutments were stored for overseas shipment, they became available as a regular property class (Class 14, augmented by plumbing fixtures, Class 29, and electrical supplies, Class 08-B) subject to requisition in the normal manner. Orders for this type of equipment were later increased, as more Air Depot Groups and, subsequently, Service Groups were activated. But by that time this equipment was a matter of routine supply both in this country and overseas; it was no longer acquired as a special facility.

### Organizational Structure

At the beginning of the expansion program the basic component units of Air Depot Groups were still the three Air Corps squadrons: Hq and Hq Squadron, the Supply Squadron, and the Repair Squadron. The latter two were responsible for the operation of three or four Mobile Units, which maintained contact with the Distribution Points supplying dispersed combat squadrons. The Mobile Units did not, however, furnish personnel to man Distribution Points; nor did they furnish skilled repair experts or equipment to accomplish third echelon work at dispersed airdromes. Their function was primarily one of supply.

At first there were no units from other arms and services in Air Depot Groups. But on 2 January 1942 Colonel Lester T. Miller, Chief of Staff of the Air Service Command, requested "that arrangements be made for the troops of other arms and services to be trained and assigned in order to bring the forty Air Depot Groups up to the strength necessary for their proper functioning." [24] This request was immediately granted by the War Department, and authority was issued at various times prior to May 1942 for the activation of other arms and services units, including several units, such as Engineer Companies, which were attached to Air Depot Groups for but brief periods.

In May 1942 the Air Service Command issued several small training manuals, setting forth the functions of each part of an Air Depot Group.[25] One manual was published for each of the three Air Corps squadrons, and one for other arms and services. (See Fig. II.) The following units were listed:

Chemical Platoon
Ordnance Company, Air Base, less Ammunition Section
Ordnance Company, Avn. (Bombardment, Pursuit, or Observation)
Signal Platoon
Quartermaster Platoon
Quartermaster Company, Truck
Quartermaster Platoon, Light Maintenance
Finance Detachment
Medical Detachment [26]

On 1 July 1942 the structure of Air Depot Groups was greatly changed. On that date new T/O's were issued, not only for the three Air Corps squadrons, but also for the others. Several new units appeared which had not been used before.[27] Among these were: Ordnance Companies, Medium Maintenance (Avn) (Q)—called "Q Companies" because of their previous affiliation with the Quartermaster Corps—and Signal Companies, Depot (Avn). Headquarters Squadrons were also reorganized to accommodate "special staff" sections composed of other arms and services personnel.

The organization of Air Depot Groups, as set forth in these new T/O's, was significant for two reasons: first, groups 41 to 80 were organized

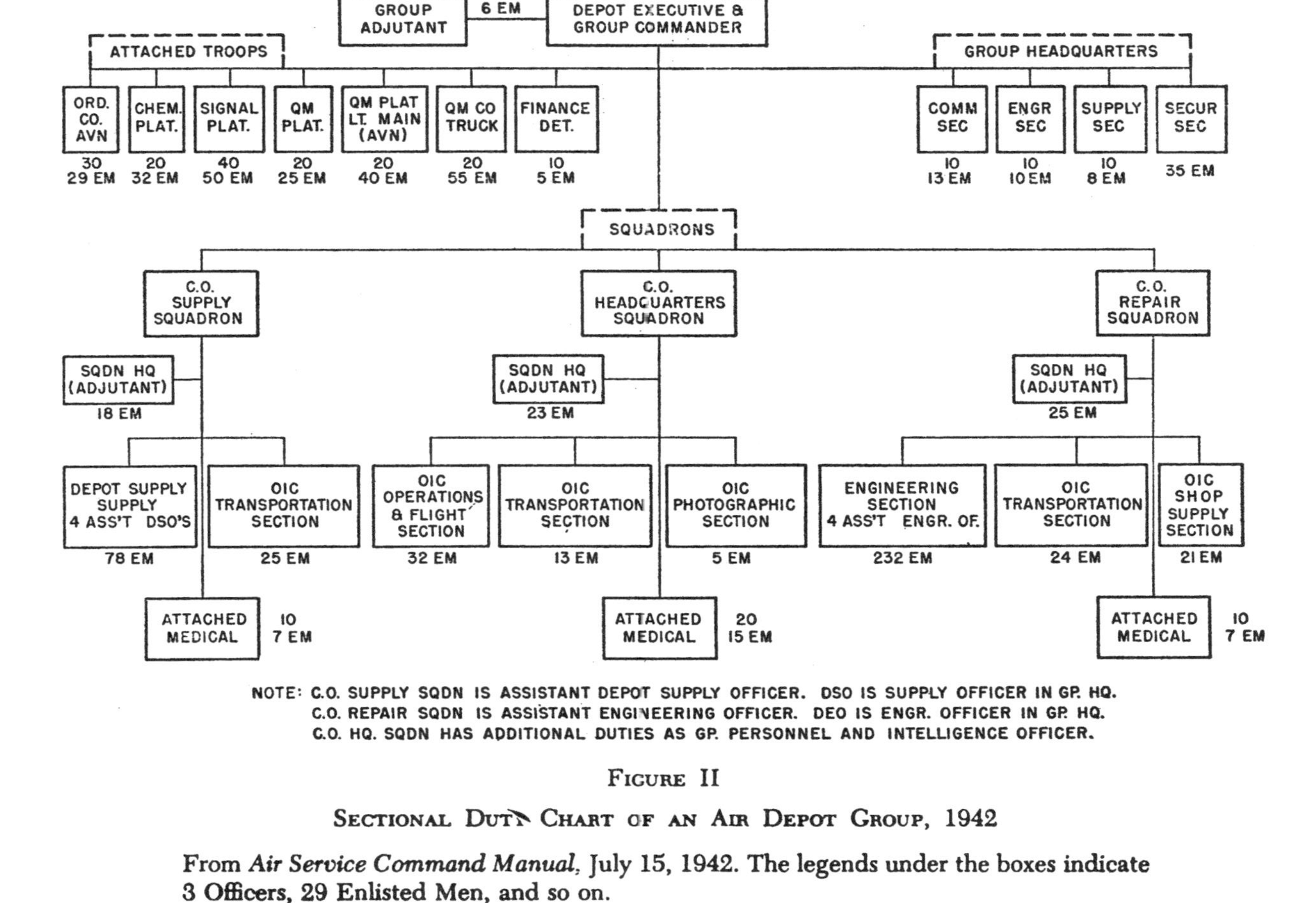


FIGURE II

SECTIONAL DUTY CHART OF AN AIR DEPOT GROUP, 1942

From *Air Service Command Manual*, July 15, 1942. The legends under the boxes indicate 3 Officers, 29 Enlisted Men, and so on.

and staffed as therein prescribed, without necessity for reorganization; second, the framework of the Army Air Forces' logistical system, as later set forth in AAF Regulation No. 65-1, 14 August 1942, was initially established in these T/O's.

## SOURCES OF ENLISTED PERSONNEL

At the beginning of the Eighty Group Program, military personnel were not, in fact, arriving at Air Service Command training areas as rapidly as had been expected. Several plans were advanced to improve this situation, but none proved effective. Perhaps the most ambitious was that submitted by G-5 (Plans) of the Air Service Command to G-3 (Training and Operations) on 17 February 1942:

> In order to obtain qualified personnel for Air Depot Groups—the prerequisite requirements of personnel [should be] made available to the draft board or to the organization that provides the assignments for draftee personnel, or by having personnel at the factories, subject to the draft, check in with the Air Corps representatives who in turn will coordinate his recommendations with the draft board.[28]

During the month of January only two Air Service Command training stations received enlisted personnel from Reception Centers: 1,295 were sent to Mobile Air Depot, and 370 to Herbert Smart Airport in Macon, Georgia. Enlisted men replaced by civilians upon the establishment of sub-depots in the same month were few; [29] the recruitment and training of civilians was a slow and tedious process. By April enlisted personnel had been sent to control depots as indicated in Table 4.[30]

TABLE 4

AIR DEPOT GROUP PERSONNEL AT CONTROL DEPOTS, APRIL 1942

| *Station* | *Number of Men* |
|---|---|
| 1. Fairfield Air Depot (Ohio) | 2,143 |
| 2. San Antonio Air Depot (Texas) | 1,931 |
| 3. Sacramento Air Depot (California) | 1,404 |
| 4. Ogden Air Depot (Utah) | 444 |
| 5. Mobile Air Depot (Alabama) | 3,629 |
| 6. Warner Robins Air Depot (Georgia) | 599 |
| 7. Rome Air Depot (New York) | 507 |
| 8. Oklahoma City Air Depot (Oklahoma) | 524 |
| 9. San Bernardino Air Depot (California) | 551 |

To stabilize the flow of recruits, various systems of reporting and requisitioning were introduced. The Maintenance Command had not been concerned in 1941 with personnel of other arms and services. Air Corps recruits had been requisitioned on WD AGO Form 210, which had been forwarded to the appropriate Corps Area, through the Chief of the Maintenance Command, one month before the necessary housing was expected to be ready.[31] Regular Army, National Guard, and Air Corps personnel available to the Chief of the Maintenance Command through recruitment at Patterson Field, Ohio, received orders from any one of three sources: the Maintenance Command, the Office of the Chief of Air Corps, and The Adjutant General of the War Department. This system, naturally, was considered confusing and inadequate, but the Army Air Forces had no standard procedure for handling military personnel of all categories.

With the coming of the expansion program, and the suspension shortly after Pearl Harbor of the provisions of the Draft Act which called for a 12-month period of military service,[32] a new system was introduced, which illustrates the trend of administration and helps to explain the conditions under which the training program proceeded. As a beginning, AAF Form 127 was established as a means of requisitioning Air Corps Personnel, and AAF Form 128 for requisitioning personnel of attached services.[33] Effective 27 February 1942 weekly reports were prepared simultaneously by all Army Air Force organizations. Yet requests for Air Corps personnel were sent to the Technical Training Command, while requests for personnel of other arms and services were sent through the Army Air Forces to Branch Replacement Training Centers, Branch Immaterial Replacement Training Centers, or Reception Centers.[34] In other words, there was still no uniform, or consolidated, system.

By September 1942, however, all air arm personnel were requisitioned directly from the Army Air Forces.[35] Thereafter both Air Corps and other arms and services personnel were reported to Headquarters, Army Air Forces,* which "allocated to the Air Forces and similar Commands the enlisted men scheduled to graduate from the Technical Schools." [36] This mild administrative reform was the beginning of a process of centralization which was not to end until almost every decision connected with the whole program was cleared through a single headquarters.

Personnel were allotted to the Air Service Command in bulk. The

* For this purpose AAF Forms 127A (which superseded Form 127) and 127B (which superseded Form 128) were used.

Enlisted Personnel Assignment Unit * made specific assignments to depots, bases, or groups. But the assignment system was unsatisfactory; too much time elapsed between the original report of a shortage and the actual arrival of an assigned replacement.[37] Shortage reports were often 15 days old at the time of receipt. It was then necessary to use these reports until the next report was received 15 days later, which caused a time lag of 30 days. Assignment was made 20 days before the enlisted man graduated from school, and allowing 10 more days for headquarters to issue travel orders, and for the individual to move, an additional 30 days was consumed, bringing the total to 60 days between the original report and the time the soldier actually arrived at his new station. It was partly to correct this situation that Replacement Depots were authorized for the Air Service Command in 1943.

The chief weakness in the methods of 1942, however, was that there was little assurance that personnel would be forthcoming merely because they had been requisitioned, or that they would be allocated by command in accordance with predetermined, or predictable, priorities. As a result, many units were activated with but one man assigned to maintain administrative records. And worse—in others 15 percent to 30 percent strength was available, but training could not be commenced until the full complement was on hand. (In some cases the men got married and raised families while they waited.) Indeed, there was never a time in 1942 when sufficient numbers were available throughout the training period for any given group. The problem was constant. It was summarized a year later by Brigadier General Junius W. Jones, the Air Inspector, in an inspection report on the Air Service Command for the Commanding General, Army Air Forces, as follows:

> Two mistakes were made . . . one by this headquarters [AAF] and one by the Air Service Command. The mistake of this headquarters was in activating the units with insufficient personnel, relying upon the Form 127 to secure the personnel which was short. The mistake of the Air Service Command was in its failure to give the officers in command of the organizations an estimate as to the date when personnel might be expected, so that such officers could make some sort of plans, or, in the alternative, to reduce the component in activated squadrons to one man so that numbers of men would not be held up in idleness and without organizational training.[38]

* Of the Enlisted Branch, Military Personnel Section, Personnel and Training Division of the Air Service Command.

## Sources of Officer Personnel

The shortage of officers in the Air Service Command may be illustrated by the predicament of the 24th and 25th Air Depot Groups at Brookley Field, Mobile, Alabama, in April 1942. The majority of officers had reported directly from civilian life with no previous military or technical training. The enlisted men also had received little or no basic military training prior to reporting to this station, and there were no non-commissioned officers of the first or second grade in either group. Moreover, there were no enlisted men in the entire command available for promotion to these grades. "Both groups were poorly prepared for field duty," wrote Colonel Lester T. Miller, "due to lack of experienced officers, senior non-coms, and sufficient time to complete their organization and training." [39]

The problem of organizing Air Depot Groups was complicated by the fact that enlisted men from other arms and services were assigned to the Air Service Command without officers. In March 1942, Colonel Miller reported that "men are being furnished to us by other branches, but they are coming to us untrained and unequipped, and not formed into any organization whatsoever. Apparently, we are supposed to train them and equip them and furnish them with commissioned officers." [40] Since responsibility for training devolved upon officers, the solution of this problem became a matter of importance.

All branches of the Regular Army had maintained a Reserve Corps for years, which had been increased by the various ROTC, CMTC, and National Guard units throughout the country. The ROTC and CMTC had been training officer personnel for Infantry, Artillery, Cavalry, and other arms of the service, but not for the Air Corps.[41] Thus most branches had a substantial reserve of trained officers with the exception of the Air Corps, in which the reserve was insignificant. Although provision was made for a few non-flying officers in the Air Corps Reserve,[42] the majority of these officers were rated flyers. The few non-flyers were mostly ex-flyers, who had been suspended from flying status. Specialist Reserve officers were rare in the Air Corps,[43] and were chiefly to be found in Washington.[44]

In 1941 the A-1 (Personnel) of the Maintenance Command had been asked to furnish the Air Staff a description of all jobs performed by officer personnel in the Air Service Command, especially in tactical units. Early in 1942 determination was made by the Air Staff that the Air Service Command required specialists, other than flying officers, in 35 different fields.[45] Officers were obtained for these 35 specialties by direct commission, both from civilian life and from the ranks. Applications were handled by the Appointment and Procurement Division, Air Staff, A-1 (Personnel), which maintained liaison in Washington with the several Air Forces and Commands.[46] In cases where applicants had formerly been employed by Federal Agencies, prior approval of release certificates was obtained from the heads of the agencies in question.[47] Assignment, wherever possible, was at some distance from the applicant's home to avoid criticism of his engaging in private business on a part-time basis.[48] Commissions from the ranks were frozen in September 1942;[49] commissions from civilian life were terminated temporarily in June 1942.[50]

Officers for other arms and services units were requisitioned in 1942 from The Adjutant General of the War Department. These requisitions were often submitted through the Corps Areas (Service Commands). AAF Form 128, previously mentioned in connection with enlisted men, was not generally used for requisitioning officers, although regulations provided that it should be. Instead, informal letters of requisition were submitted with lists that were made up on forms prepared at Headquarters, Air Service Command. A typical letter of requisition, dated 4 April 1942, is quoted in part below:

> The attached inclosure showing officer requirements for the Signal Corps, Signal Platoons, for Air Depot Groups as indicated, is submitted herewith with dates and stations to which these officers should be ordered to report for duty.
>
> In most cases these dates will allow the officers to report three days in advance of the time designated for the enlisted personnel of the Arms and Services to join their respective Air Depot Groups. Copy of the training schedule for Air Depot Groups for the calendar year, 1942, is attached.
>
> It is urgently requested that officer personnel be assigned to the Arms and Services for the 2nd Air Depot Group at the earliest practical date, and that the 6th, 7th, 8th, 9th, 10th, 13th, 24th and 25th Air Depot Groups be next in the order named in the assignment of Arms and Services officer personnel.[51]

In September 1942 the policy of requisitioning officers and enlisted men, both Air Corps[52] and other arms and services, was standardized with the introduction of AAF Forms 127A and B.[53] Requisitions made directly on The Adjutant General were discontinued, except in the case of very high ranking officers.[54] Headquarters, Army Air Forces, became responsible for the assignment and transfer of all officers attached to service units.

Officers graduating from OCS and OTS at Miami Beach, Florida, were also assigned by Headquarters, Army Air Forces, although for a short time requests had been submitted by all sorts of AAF activities directly to those schools.[55] When the Miami schools were established in February 1942,[56] they became the chief source of Air Corps officers for all commands and air forces. At that time large-scale transfers from other branches of the service, and reassignment of graduates of Branch Immaterial (Army Administration) OCS's, had not yet begun. The Army Air Forces and the Air Service Command were obliged to train their own personnel. Officers temporarily assigned to air force activities from the other arms were recalled to their respective branches. For example, Coast Artillery officers were released by the Army Air Forces in August.[57] In November 1942 the Air Service Command opened its own school with the activation of the Third Air Service Command and Staff School at Warner Robins, Georgia.[58]

## Group Functions, Including Engine Overhaul

The mission of Air Depot Groups in the first part of 1942 was limited to second and third echelon work for three air force tactical groups.[59] This work included the supply, maintenance, repair, and salvage of military aircraft—all that could then be attempted in the theaters of operations. It also included services to the troops by the several Air Corps and others arms and services units attached to the group.

The organizational equipment of an Air Depot Group was that used by a Base Depot. It approximated half the equipment used by the Middletown Air Depot in Pennsylvania prior to the period of expansion.[60] The equipment of one Repair Squadron was sufficient to keep the personnel of three such squadrons busy eight hours a day, if used continuously, and at Base Depots, such as the Warton Air Depot in the United Kingdom, it was used in just this manner. When consolidation was undertaken, two or three Air Depot Groups were assigned by the Theater Commander to a single installation.[61] At that time the Air Service Command had no other tactical organization with which to implement an Air Depot.

Under these conditions it was difficult to obtain the proper balance between adequacy of equipment and mobility. "On a recent trip to Charleston," wrote Colonel T. J. Hanley, Jr., Assistant Chief of Air Staff, A-4 (Supply), "I looked over the equipment of the Depot Group and was surprised to find that it amounted to so much. The so-called 'portable' hangars and shops can hardly be considered mobile, having been once erected." [62] This situation was admitted by the Air Service Command. The difficulty of moving an Air Depot Group had been amply demonstrated by the 4th Air Depot Group in the Louisiana-Carolina Maneuvers. And the 4th had been traveling light! Nevertheless, Air Depot Groups were directed to maintain the highest degree of mobility and flexibility consistent with the equipment and supplies assigned to them.

One type of equipment which was not assigned to Air Depot Groups was aircraft. Transportation was accomplished by truck, augmented by railroad or shipping facilities. In the few cases in which supply by air was necessary, airplanes were provided by the Air Transport Command, or by a combat commander. In the theater of operations, Air Depot Groups maintained an installation complete with landing strip or landing field, and depot facilities, but the airplanes were all borrowed.[63]

The original intent was not to include engine overhaul. This class of work was undertaken by a few groups, however, as a result of a directive issued by Major General Henry J. F. Miller at the end of 1941.[64] In January 1942 procurement of engine overhaul equipment was authorized for "one-half the total of Air Depot Groups to be activated." [65] On the 19th of that month Colonel Hicks reported that "to date only 15 equipment sets for the overhaul of engines and engine accessories have been placed under procurement . . . but immediate action will be taken to initiate a procurement program for an additional 60 equipment sets." [66]

Since Air Depot Groups were destined for service in distant countries, and since equipment and supplies sent overseas would use critical shipping space, the point at issue was largely a matter of tonnage and cubage. Would the equipment necessary to overhaul 1,000 engines and accessories per year occupy more or less space on board ship than 1,000 additional engines and accessories? And which loss could the nation afford more readily? Colonel T. J. Hanley, Jr., Assistant Chief of the Air Staff, A-4 (Supply), wrote: "I believe that a study will show that the supplies, equipment, and machinery necessary to overhaul 1,000 engines a year will greatly exceed the space necessary to transport 1,000 new engines from the United States to the places where they are needed."[67] It was the opinion of the Air Service Command, however, that "the equipment necessary to set up an engine overhaul will be less than 50% of the space required to transport 1,000 tactical engines."[68] Also, engine overhaul, in close proximity to the using agencies, would reduce the total number of engines required and cut down the number of spare parts and accessories that would have to be manufactured. Moreover, this service at the scene of operations would eliminate the perils of a round-trip voyage across submarine infested seas; once equipment arrived at a theater of operations, it would not have to be sent home for repairs. For these reasons tentative arrangements were made for attaching engine overhaul facilities to Air Depot Groups.

### Equipping Air Depot Groups[69]

Air Depot Groups were equipped by means of Organizational Equipment Lists (OEL's), which had been initiated early in 1941, and were printed and mailed twice a year for purposes of inventory and requisition.[70] Certain sections of these OEL's were withdrawn by the Budget Office, and forwarded directly to the Field Service Section (later redesignated Air Force Section) at Wright Field, Ohio, whence the items were delivered automatically. Other parts of the OEL's were mailed to the supply officers of Air Depot Groups, who filled in the items authorized by T/BA in the routine manner.[71] A few sections of the OEL were not printed at all, for they applied to aircraft and engine parts and spares, which were not issued to Air Depot Groups.

Whenever equipment was to be shipped to any squadron of an Air Depot Group, it was addressed to the supply officer of the group, rather than to the supply officer of the station at which the group was located.[72] It had been found that whenever shipments were made to station supply officers, unnecessary paper work resulted, and the groups were delayed in receiving their equipment. Since Air Depot Groups carried their own accountability, both for organizational equipment and for maintenance parts and spares, supplies of all kinds could be sent to them directly from any air depot in the country.

The normal arrangement was to assign an Air Depot Group to one of the continental air depots for purposes of supply. This was usually, but not always, the depot at which the group was located. After the OEL's had been processed by this depot, each squadron of the group reported to the Chief, Field Services, as to what items had been received.[73] Reports were submitted on Mondays by teletype addressed to the attention of the Equipment Supply Group (later "Section"). Only items of Air Corps supply, exclusive of those manufactured locally, were so reported.

Separate reports were submitted monthly on items of other arms and services supply furnished to the Air Corps units.[74] This equipment was requisitioned in 1942 through the office of the appropriate arm or service at the station at which the group received its equipment. If the arm or service was not represented at that station, requisition was made through the Corps Areas. Units of the branches attached to Air Depot Groups utilized the same channels as Air Corps units, but their system of reporting was determined by the branch to which they belonged.

Overhaul equipment for engines and engine accessories was not assigned to Air Depot Groups until it was definitely known whether or not the group would be employed in this work. Often it was not known until the groups were functioning overseas. Nevertheless, complete sets of overhaul equipment were available at certain of the control depots, and were shipped to the theaters, or to Ports of Embarkation, upon authorization by the Air Force Section of the Air Service Command.

Aircraft spares and maintenance parts were not authorized for Air Depot Groups in the United States unless the groups were specifically designated as service organizations for tactical units in maneuvers, or at training centers. On receipt of warning orders for shipment overseas, however, Air Depot Groups were entitled not only to submit shortage lists to complete their authorized equipment, but also to receive an initial 90-day level of maintenance parts and spares to be used on arrival at the theater of operations.[75] Thereafter, they requisitioned through the Task Force Commander.[76]

★★★ 4 ★★★

# BEGINNING OF THE TRAINING PROGRAM

## Early Efforts to Standardize Procedures

The training of personnel for Air Depot Groups has always been a difficult task—more complicated and technical by far than the training of most military units. As the nation's industry went into high gear producing aircraft, engines, instruments, and machinery, the air forces were hard pressed to train people to service this equipment. Gigantic industrial establishments, typified by the Willow Run Plant of the Ford Motor Company, were planned, built, tooled, and put into operation in a period of months. The Air Service Command found itself faced, overnight, with the problem of converting automobile mechanics, stock brokers, farmers, and grade school boys into skilled aircraft technicians. A realization of the magnitude of this job dawned on personnel of all echelons of the command alike.

Explaining the engineering features of military aircraft was not enough in itself; work in a tactical service unit was more than an eight hour a day job. In a sense, there was no clear line between the "training program" and any other program. Everything that happened to a group committed for overseas duty was part of its training. Often the best preparation was a move from one location to another in bewildering and annoying circumstances, or the setting up of hangars, warehouses, and housing facilities at a newly established airfield, or the servicing of air force units engaged in a tactical function, such as anti-submarine patrol. A group with experienced leaders, wise in the ways of the army and aware of the multiplicity of the tasks to be performed, became relatively efficient, regardless of formal training; a group with poor leaders became a poor group. Both types were shipped overseas.

Training officers at the headquarters responsible for this program knew that all types of practical experience would serve the groups in good stead overseas. Decentralized activity was encouraged.[1] Initiative on the part of squadron, group, and depot commanders was welcomed; individual leaders were given a free hand. "Tanforan is yours" was the only field order that Brigadier General Edwin S. Perrin gave to Lieutenant Colonel Louis A. Merillat, Jr., in directing the reorganization of administration and training at that West Coast base.[2]

Detailed directives were not issued until after acceptable procedures had been evolved through practice. Inspection parties came from higher headquarters as much to learn what could be done in the field as to ascertain whether or not the policies of the command were being adhered to. Whenever manuals, directives, or training aids in a particular locality lent themselves to general application, they were standardized and incorporated into the training doctrine of the command.[3] Outstanding leaders and instructors in lower commands were often transferred to higher levels to carry out their policies on a larger scale.

After the expansion began, reservoirs of technically trained enlisted men were quickly exhausted. The first four Air Depot Groups in 1941 had consisted largely of Old Army men from Materiel Squadrons, Transport Squadrons and sub-depots, augmented by civilians recruited for the groups at the depots. These men were already trained technicians, who needed only brief periods of on-the-job work with Air Depot Group equipment. For a short time in 1941, therefore, Air Depot Groups had sizable complements of picked men, but this brief period was quickly forgotten. Even before the declaration of war, large numbers of green inductees were required to fill unit T/O's, and from then until the end of the war service units were always in need of training.

In 1941 unit training was conducted exclusively at the depots. Responsibility for instruction, technical and military, belonged to group commanders subject to the supervision of depot commanders, under the Commanding General of the Maintenance Command (Air Service Command). Even after the four Air Service Area Commands were set up shortly after Pearl Harbor[4] responsibility for the training of Air Depot Groups remained largely with the depots at which the groups were stationed.

As time went on, however, Headquarters, Air Service Command, exerted an increasing influence on the development of its service units. This influence was exercised in the formulation of policies establishing the framework within which lower echelons operated. Aside from such preliminary matters as the selection of training sites and the screening and placement of officers and enlisted men, it made numerous and far-reaching decisions on training policy. One of the most important was the establishment of technical schools at Air Service Command depots and sub-depots to supplement the training at Air Corps Technical Schools (later Technical Training Command Schools). This decision in turn led to the utilization of civilian schools and colleges [5] and to the procurement by this command [6] of a great quantity of training aids and equipment.[*]

The over-all plan of the Army Air Forces had been to have the Technical Training Command accomplish individual technical training and basic military training, and to have the Air Service Command accomplish unit training, by squadrons or companies, and combined training, bringing together the various squadrons of a group for the performance of a tactical mission.[7] This arrangement did not work out, because the bulk of the personnel received for unit and combined training had not previously received sufficient individual training. The Air Service Command thus had no alternative but to assume this responsibility, and to sacrifice the unit and combined training originally assigned to it.[8] It was the opinion of the Air Service Command that the instruction offered by the Technical Training Command was helpful, but that actually in many specialties it was not sufficiently advanced to include third echelon work. Moreover, there was a surprising number of cases in which the Air Service Command received personnel directly from Reception Centers, fresh from civilian life.[9]

Other reasons for the providing of technical training by the Air Service Command were secondary to its necessity. It was true that the depots had much of the required equipment, but the scope of an air depot, in

[*] Training aids were defined in AAF Reg. No. 50-19 (several dates) to include the following:

(a) Literature and informational documents.
(b) Recordings, metal wire, film and similar materials processed for reproduction of sound.
(c) Motion picture films, film strips, film slides and target films.
(d) Posters, graphic material and other visual aids.
(e) Training devices, including ground instructional equipment (except for complete aircraft) as defined in AAF Regulation 65–53, synthetic trainers, demonstration devices, mock-ups used for training purposes including those used in mobile training units, training devices incorporating operational equipment, adapters and accessories for training turrets, radio and radar trainers, photographic and personal equipment trainers, all targets and any modification or combination of the above.

both engineering and supply, was vastly greater than that of an Air Depot Group. Operating a 16 inch lathe at a depot was not the same as operating a 6 inch lathe in the field. Nor were the procedures of supply that were applicable to 15,000 civilians the same as those which an Air Depot Group could undertake at an isolated air base. The problems of these occupations were generally related, nevertheless, and this argument was often used in 1943 to justify the existence of unauthorized schools. But the reason for such schools, originally, was to train men who had been turned over to the Air Service Command unable to perform the functions for which they were required.[10]

Because of the necessity for continuing basic technical and military training, many groups were deprived of unit and combined training. Theoretically, individual training of all enlisted men had been completed before the unit training program of the Air Service Command began. Whatever time was devoted by the Air Service Command to individual training, therefore, was stolen from unit training, as envisaged in the original plan, with the result that both unit and combined training were often altogether dispensed with.

There was a further complication in that officers had to be trained to assume their group responsibilities while the units were being manned.[11] Consequently, they had no opportunity to become acquainted with their enlisted men, or to take hold of the administration of their units, and many important decisions, including the selection of officer candidates, necessarily devolved upon high ranking non-coms.[12] This situation, and the lack of training on the part of the officers, created a serious morale problem.

In December 1941, in anticipation of the expansion of training activities entailed in the Forty, and later in the Eighty, Group Program, Colonel Lester T. Miller, Chief of Staff of the Air Service Command, wrote to his G-3 (Operations and Training) directing that letters be issued to all depot commanders, instructing them to start schools for supply and engineering officers. Manuals were also to be written at the depots covering all phases of the operation of the new groups. Moreover, standardized training plans were to be prepared by the Air Service Command. immediately. "These training plans should be issued in the form of directives to the group commanders," he wrote. "It must be assumed that the training period will not be long, that the officers will be inexperienced and, therefore, the directives should be very detailed." [13]

The first training directive of general application issued by Headquarters, Air Service Command, was published on 17 February 1942; it was entitled "Master Training Program for Air Depot Groups." [14] In this directive, the responsibility of the command was broken down into the following categories:

| Basic or Individual Training | Advanced or Unit Training |
|---|---|
| Technical | Section |
| Military | Squadron |
| | Group |

Provision was made for local variations. Group commanders were informed that they were responsible, under depot commanders, for the preparation of programs and schedules to effect the proper sequence and uniformity in the training of their units.[15] They were to adapt the Master Training Program to their own needs, with consideration for the facilities and equipment at their disposal, and to submit their plans and schedules to Headquarters, Air Service Command, for approval. They were also to have ready for inspection at any time training progress reports for both units and individuals.

Almost immediately it became apparent that correct channels of communication were not being followed. Plans and schedules prepared by group commanders were submitted in many cases directly to Headquarters, Air Service Command, by-passing the four Air Service Area Commands (which were responsible to the Air Service Command) and sometimes even the control depots.[16] A subsequent directive ordering semi-monthly "Squadron Training Progress Reports" repeated the same mistake. Reports were to be forwarded so as to reach Patterson Field, Ohio, not later than the 5th and the 20th of each month.[17] In the Training Section of at least one of the four areas, however, these reports were not processed as a matter of normal routine until October 1942.[18] Yet the information they contained was of increasing importance in the development of sound doctrines throughout the command.

Training standards for enlisted men were to be indicated by proficiencies specified in AR 615–26, the basic army regulation prescribing technical qualifications for the several specialties in Air Depot Group T/O's. Attention was directed [19] to appropriate Technical Orders, Materiel Division Manuals, Manufacturers Manuals and those Field Manuals and Technical Manuals that were pertinent to the Army Air Forces.

A list of the two last-named types of publications used in the training of Air Depot Groups in February 1942 is given in Table 5.

TABLE 5

PUBLICATIONS FOR AIR DEPOT GROUPS, FEBRUARY 1942

| No. | Title |
|---|---|
| **FIELD MANUALS** | |
| 21–10 | Sanitation and First Aid |
| 21–15 | Equipment, Clothing, and Tent Pitching |
| 21–20 | Physical Training |
| 21–30 | Conventional Signs, Military Symbols, and Abbreviations |
| 21–40 | Defense Against Chemical Attack |
| 21–50 | Military Courtesy and Discipline |
| 21–100 | Soldier's Handbook |
| 23–5 | U.S. Rifle, Caliber .30, M1 * |
| 23–10 | U.S. Rifle, Caliber .30, M1903 * |
| 25–10 | Motor Transport |
| 26–5 | Interior Guard Duty |
| 27–15 | Military Law, Domestic Disturbances |
| **TECHNICAL MANUALS** | |
| 1–505 | Aircraft Cameras (Machine Gun) |
| 9–980 | Bombs for Aircraft |
| 10–545 | Motor Transport Inspections |
| 1–405 | Aircraft Engines |
| 1–406 | Aircraft Electrical Systems |
| 1–407 | Aircraft Induction, Fuel and Oil Systems |
| 1–408 | Aircraft Engine Operation and Test |
| 1–409 | Aircraft Armament and Pyrotechnics |
| 1–410 | Airplane Structures |
| 1–411 | Airplane Hydraulic Systems and Misc Equipment |
| 1–412 | Aircraft Propellers |
| 1–413 | Aircraft Instruments |
| 1–415 | Airplane Inspection Guide |
| 1–420 | Lathes |
| 1–421 | Milling Machines, Shapers, and Planers |
| 1–422 | Grinding Machines |
| 1–430 | Welding |
| 1–435 | Aircraft Sheet Metal Work |

* Depending on type of rifle issued.

| | |
|---|---|
| 1–440 | Parachutes, Aircraft Fabrics, and Clothing |
| 1–455 | Electrical Fundamentals |
| 1–460 | Radiotelephone Procedure, Air Corps |
| 1–490 | Electrical Armament Controls |
| 1–495 | Aircraft Machine-gun Sights |
| 1–500 | Bomb Racks, Tow Target Equipment, and Flare Racks |

On 8 April 1942 eleven Station Complements were activated by the Air Service Command, and shortly thereafter several others were added.[20] Located at stations training Air Depot Groups, they were created largely to provide administrative and supply facilities, including those of other arms and services, for troops of trainee tactical units. All of the bases at which Station Complements were activated were designated as Air Depot Training Stations. Most of them were at Air Depots; five of the most important were not. Thus Air Depot Training Stations were activated at Albuquerque, New Mexico; [21] Stinson Field, San Antonio, Texas; [22] Ten Mile Station, Charleston, South Carolina; [23] State Fairgrounds, Springfield, Illinois; [24] and Hensley Field, Texas.[25]

By the beginning of July 1942 twenty-two Air Depot Training Stations were busy obtaining equipment and setting up facilities for the training of the various units of Air Depot Groups.[26] The duplication of effort involved in this dispersion led several responsible officers to protest its inefficiency to Headquarters, Air Service Command. On 10 September 1942, for example, Colonel Leo H. Dawson wrote from Albuquerque, New Mexico, to the Commanding General, Air Service Command, attention of Brigadier General McMullen:

It is therefore recommended that the entire training program of Air Depot Groups be rearranged so as to utilize the best facilities, best schools, the best instructors, more comprehensively, and to eliminate the more unfavorable training stations. . . . I talked this plan over with Jimmy Spry [Major Spry, Hq ASC] who is in favor of it, as it will save putting machine tools in many places. This will give more tools for shipment overseas.[27]

The reply, written on 14 September 1942 by Colonel James D. Givens, Training and Operations Officer, it was addressed to Colonel Dawson at the Hilton Hotel, Albuquerque, New Mexico:

Your plan is very similar to the directive we received from General McMullen which directed that we concentrate training of Air Depot Groups at not to exceed five locations, and suggested that we use Albuquerque, Stinson, New

Orleans, Wilmington, and Charleston for this purpose. I invited McMullen's attention to the fact that General Henry F. Miller himself * had directed that mobilization type construction be built at each one of our eleven Control Depots. All this construction is now nearing completion, and to avoid national scandal, it will be necessary for us to keep this housing occupied.[28]

The opinion expressed by Colonel Givens seems to have prevailed, although there may have been other reasons for the establishment of so many training stations. At any rate, Air Depot Group training was not consolidated at favorable locations for many months. Thus, extra equipment had to be procured, and detachments, sections, and units were dispatched from one end of the country to the other in search of specialized instruction. Squadrons and groups which had exhausted the training possibilities of one location were moved to another, where new types of facilities and equipment were available. The task of supplying these peripatetic units with their organizational equipment and even the barest of necessities became harder as the number of groups increased. "Starting December 9th," wrote Colonel D. R. Stinson, Assistant Chief, Supply and Services Division, Office Assistant Chief, Air Staff, Materiel, Maintenance and Distribution, "practically every squadron in the United States was moved from some [place] to another place; and from that day to this [1943] their original equipment has never caught up with them. . . . Operations Officers appear to be entirely oblivious to the problem of packing and moving supplies." [29] This criticism was not leveled at the training program of the Air Service Command in particular, but it certainly described conditions prevalent in the training of Air Depot Groups.

When Station Complements were organized, they were charged with coordinating the training activities of the post in the name of the depot commander. The method of training was "by exercise of normal command in administration and technical operations for each individual and each echelon." [30] Operations were to be executed step by step, but they were to be sufficiently diversified to keep each unit in a state of readiness for immediate employment.

The training was a combination of individual and unit (squadron or

* As a matter of fact, the directive to set up Mobile Air Depot Groups at each of the eleven depots had come not only from General Miller, but from General Arnold. See teletype E-273, 29 Sept 1941, and 2nd Ind: Lt. Col. Merwin E. Gross, Asst, Exec., Office Chief of Air Corps, to Chief of the Army Air Forces, 5 Nov 1941, in TSAGD, Microfilm Unit, 1941 Correspondence, Reel 13, Item 11.

company) training. It was also a combination of technical and military training.[31] Post schools were set up, and courses were given by a faculty consisting of a few full-time instructors and many other officers recruited for lectures and demonstrations, often on short notice, from the shops of the depot proper, and from various spots in the cantonment or troop training areas. Emphasis was placed on individual technical training, with as much on-the-job work as possible, but military training was squeezed in when the troops were not attending these courses. In many cases the importance of continuous instruction in technical subjects was not sufficiently appreciated by the officers and warrant officers in charge of administration. Too frequently soldiers exploring the intricacies of radar or aircraft instruments were yanked out for "shots," or lectures on Sex Morality and the Articles of War.[32]

A typical plan was prepared by the Fairfield Air Depot, consisting of a cycle of six-week periods to provide a steady flow of processed and trained personnel. (See Fig. III.) It was never possible, however, to have all personnel of a given squadron trained in one place simultaneously, for many of the men were away on temporary duty at contract schools during the entire time their unit was in this country.[33] So, while some individuals were trained by Station Complements, others were dispatched to contract schools. When the members of a squadron were finally assembled, it was often found that there was a considerable diversification in training, and that some individuals had missed out on military training altogether.[34]

## Technical Training

Technical training for Air Depot Groups was largely the training of individuals for specialized tasks in connection with the servicing of aircraft. Generally speaking, the same procedures of instruction were used as those which had been worked out in the Air Service Command's program for the training of civilians.[35] In 1941 occupational analyses had been made by several of the depots to assist in the classification and training of civilian employees, and these studies were of great value in the subsequent training of enlisted men. Occupations had been broken down into jobs, and jobs into separate operations. Such breakdowns indicated step-by-step procedures, and were simpler for both instructors and pupils to follow than Technical Orders, which had been the only standard instructional material available before the war. Training

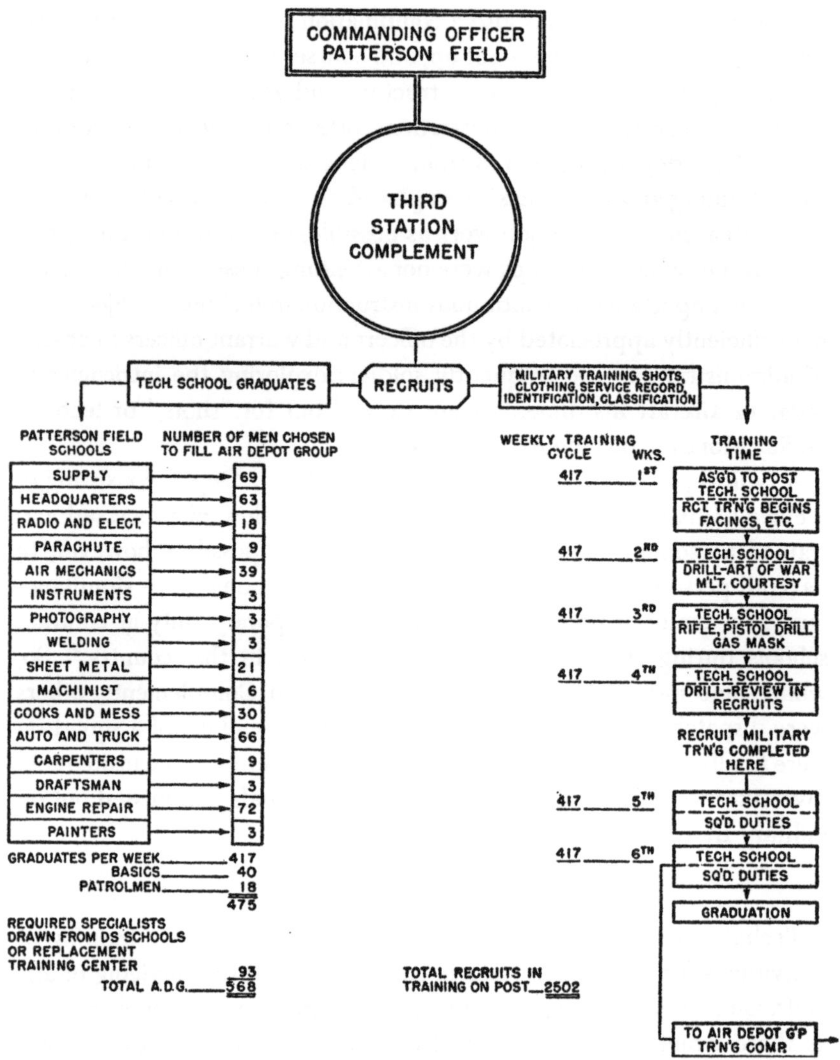


FIGURE III

SIX WEEKS' TRAINING PLAN FOR RECRUITS

aids of this sort were of particular value, as qualified instructors with vocational training experience were rare, in or out of the army.

In February 1942 it was estimated by G-3 (Training and Operations) of the Air Service Command that only 20 percent of the technical personnel required for Air Depot Groups would be forthcoming from Air

Corps Technical Schools (Technical Training Command Schools), and that the remaining 80 percent would have to be trained by the Air Service Command, for the most part at its depots and contract schools.[36] Off-reservation schools, financed by National Defense Training funds, and factory schools [37] maintained by civilian industries were also to be utilized by the Air Service Command, and were of much importance at a later date.

Nine civil schools were under contract in February 1942 to train 4,604 students—48 percent of the anticipated total.[38] Two additional contracts were negotiated shortly afterwards by the Contract Unit of the Materiel Command,[39] using Army Air Force funds, making a total of eleven contract schools in March 1942, including such well-known institutions as Parks Air College and the Boeing School of Aeronautics.[40] By June 1942 sixteen civil mechanics schools, with courses in about a dozen overhaul specialties, had become the direct responsibility of the Air Service Command.[41] Approximately 8,000 enlisted men were trained at these schools while the courses of instruction were supervised by the Air Service Command.[42]

Soldiers at contract schools were organized into Air Corps Training Detachments under commissioned officers.[43] In order to fill quotas, they were sent directly from tactical groups. Administration of the program would perhaps have been facilitated if the Air Service Command had had Reception Centers from which to draw personnel, but no such installations were authorized in 1942. The first men available for assignment to Air Depot Groups, or to units of attached arms and services, were of necessity considered a reserve pool for supplying school requirements.[44] The number of men selected was determined by the quota of technicians required in each category for groups on high priority for foreign service.

Under this system it was inevitable that a certain number of individuals selected from a given group would be sent to another group upon graduation.[45] Consequently, group commanders were reluctant to release their best men; there was a tendency to send "bolos" and hope they would not come back. This practice led the supervisors of contract schools to complain about the lack of qualifications of the men who were sent to them for training.[46] Moreover, there was increased pressure upon the Air Service Command for the establishment of Replacement Centers, so that enlisted men might be sent to schools without robbing tactical, and in some cases committed, units. A brief statement of the number of enlisted men who were sent to contract schools during the time they were under the Air Service Command is given in Table 6.

TABLE 6

COURSES AT CONTRACT SCHOOLS OF THE AIR SERVICE COMMAND *

| | | |
|---|---|---|
| Electricians, Aircraft | | |
| Aero Industries Technical Institute | 162 | |
| Embry-Riddle School of Aviation | 168 | |
| | | 330 |
| Machinists, Aircraft | | |
| National School of Aeronautics | 425 | |
| Ohio State University | 16 | |
| University of Wisconsin | 52 | |
| Williamson Free Trade School | 18 | |
| | | 511 |
| Mechanics, Aircraft | | |
| Aero Industries Technical Institute | 162 | |
| Aviation Institute of Technology | 285 | |
| Boeing School of Aeronautics | 75 | |
| Embry-Riddle School of Aviation | 100 | |
| Ohio Institute of Aeronautics | 179 | |
| Parks Air College | 248 | |
| Robertson Aviation School | 123 | |
| | | 1,172 |
| Mechanics, Electrical Instruments | | |
| American School of Aircraft Instruments | 40 | |
| Chicago School of Aircraft Instruments | 43 | |
| | | 83 |
| Mechanics, Engine | | |
| Aero Industries Technical Institute | 174 | |
| Aviation Institute of Technology | 304 | |
| Boeing School of Aeronautics | 150 | |
| Curtiss-Wright Technical Institute | 150 | |
| Embry-Riddle School of Aviation | 716 | |
| National School of Aeronautics | 469 | |
| Ohio Institute of Aeronautics | 192 | |
| Parks Air College | 275 | |
| Southwestern Institute of Technology | 125 | |

* From Statistical Control, "Training Report—Technical Training," Corrected Report for Period Ending 30 June 1942, in AFIHI archives.

| | | |
|---|---|---|
| Stewart Technical School | 478 | |
| St. Louis School of Aeronautics | 190 | |
| | | 3,223 |
| Mechanics, Gyro Instrument | | |
| American School of Aircraft Instruments | 50 | |
| Chicago School of Aircraft Instruments | 43 | |
| | | 93 |
| Mechanics, Mechanical Instrument | | |
| American School of Aircraft Instruments | 40 | |
| Chicago School of Aircraft Instruments | 43 | |
| | | 83 |
| Mechanics, Parachute | | |
| Curtiss-Wright Technical Institute | 50 | |
| | | 50 |
| Mechanics, Propeller | | |
| Curtiss-Wright Technical Institute | 24 | |
| Parks Air College | 25 | |
| | | 49 |
| Mechanics, Sheet Metal | | |
| Aero Industries Technical Institute | 363 | |
| The Anderson Organization | 499 | |
| Boeing School of Aeronautics | 45 | |
| Curtiss-Wright Technical Institute | 542 | |
| Embry-Riddle School of Aviation | 100 | |
| Ohio Institute of Aeronautics | 88 | |
| Robertson Aviation School | 60 | |
| | | 1,697 |
| Welders, Aircraft | | |
| Aero Industries Technical Institute | 165 | |
| Casey Jones School of Aeronautics (discontinued 30 June 1941) | 42 | |
| Curtiss-Wright Technical Institute | 75 | |
| Embry-Riddle School of Aviation | 100 | |
| Ohio Institute of Aeronautics | 64 | |
| Robertson Aviation School | 20 | |
| | | 466 |
| Grand Total | | 7,757 |

The success of the technical training of enlisted men during this period was more difficult to judge than that, for example, of the similar program of the Air Service Command for civilians. The morale of these men was not the object of solicitous attention. No one was interested in why they were having trouble with their work, or whether or not they desired to resign. Once their brief training was completed, off they went to parts unknown. What these men thought of their training can be learned only from conversations with the individuals concerned. Those who felt that they were getting the short end of the bargain may or may not have been in the minority. The inconclusive nature of the records on this subject is described by the historian of the Sacramento Air Depot:

> There being no central responsibility for troops, admittedly so by the Commanding General, it is not surprising to find that apparently there existed no central responsibility for the keeping of a record of their activities and of their training. Beyond many hundreds of pages of copies of Special Orders, a painstaking and prolonged search for data which could be utilized historically has revealed exceedingly little. Detailed files in the training of Air Depot and Air Service Groups existed, but it is reported, on good authority, that they were, probably all, consigned to the flames.[47]

On 1 June 1942 administration of technical training of military personnel at contract schools was turned over to the Technical Training Command; thereafter the Air Service Command was to be concerned primarily with the administration of unit and combined training. The relationship between the Technical Training Command and the Air Service Command was defined on 25 June 1942 by Headquarters, Army Air Forces, as follows:

> The Commanding General, Air Forces Technical Training Command, Knollwood Field (Southern Pines), North Carolina is responsible for, and has jurisdiction over, training of all military and civilian technical training students for the Army Air Forces except those technical students being trained in Air Service Command Depots, Sub-depots, and Mobile Depots, except as otherwise agreed to by the Technical Training Command or authorized by this Headquarters.[48]

After the transfer, contract schools, under the Technical Training Command, continued to train military personnel for the Air Service Command,[49] but there was greater necessity for technical training at depots, Air Depot Training Stations, and bases engaged in the training of Service Groups. Throughout the year 1942, indeed, there was a growing emphasis on technical courses at the post schools of the Air Service Command, for officers, enlisted men, and cadres of the two combined. When overhaul courses at contract schools were discontinued altogether by the Technical Training Command in June 1943, the Air Service Command was ready with its own individual technical training, in addition to unit and combined training.

### Military Training

The purpose of military training was to facilitate the execution of the technical duties of the group in the theaters of operations. Necessary elements of this indoctrination were the development of military bearing and courtesy, and practice in the care and display of equipment. These subjects were stressed throughout—particularly at Saturday morning inspections. Training programs at the several air depots and training stations varied, but the general outlines, as set forth in the Master Training Program,[50] were the same at all stations. Discipline and morale, it was stated, would be obtained as a result of good leadership and efficient administration. Physical hardening to ensure endurance under field conditions, and qualification in assigned weapons, would also be contributory factors in morale and efficiency.[51] An intensive training program to attain these ends was a part of the schedule of each section of an Air Depot Group.

In addition to the general military training of all personnel, specialized unit training was given to each squadron and platoon, according to its branch of service. Although instruction was offered to everyone in such subjects as defense against chemical attack, first aid, and convoy driving, the personnel of Chemical, Medical, and Quartermaster Truck units were given greater detail in their respective specialties. Thus, a Chemical Warfare officer in an Air Depot Group had a dual teaching responsibility: first, to give a general orientation in CWS subjects to all and sundry; second, to prepare the men of his own unit to be specialists capable of teaching others. In this manner both individual and unit training was accomplished.[52] The principle followed in developing groups was similar to that used elsewhere in the army in the training of regiments and divisions.

In the Air Corps in 1942, however, complications were caused by the lack of qualified officers in attached arms and services, and the inability of the branch chiefs in some cases to train and equip their own personnel. "Major Wilson has reported informally to this office," wrote Colonel Lester T. Miller in February 1942, "that the various sections of service personnel, other than Air Corps, which go to make up a complete Air Depot Group are not to be trained or equipped by their respective branch chiefs: that it is considered that the Air Service Command will train and equip this personnel after it has reported to the group." [53] He went on to say: "This office has always gone under the assumption that these sections would report trained and equipped to join an Air Depot Group a short time prior to embarkation. . . . We hold that the law requires that it is the responsibility of respective branch chiefs not only to train but to equip their personnel for such duty." [54] In spite of this protest, personnel for other arms and services units continued to report to the Air Service Command untrained, unequipped, and lacking in commissioned and non-commissioned leaders.

A shortcoming in the training of Air Corps personnel—symptomatic of a condition which scandalized Ground Force officers—was the laxity in "firing" troops with their prescribed weapons. As late as June 1942 the Ordnance Section reported to G-3 (Training and Operations) of the Air Service Command that "Reports from all Air Depot Groups from the 6th through the 47th . . . show that no training ammunition has been expended since January 1st by any of these Air Depot Groups with the exception of the 9th, 11th, 24th and 25th." [55] This situation was amplified by the Ordnance Section with the statement that "the reason in many cases is that the groups did not have ammunition on hand, but of the 13 groups which did have ammunition on hand, only 4 groups expended it, and these groups expended only a small fraction of the ammunition on hand." [56] They might have explained that few Air Depot Training Stations were fortunate enough to boast of rifle ranges.[57] In several instances troops were "fired" only after journeying to neighboring Infantry camps, as the 36th and 37th Air Depot Groups at Herbert Smart Airport journeyed to Camp Wheeler.[58]

Aside from the question of morale in this situation, there was an element of danger to the men. On 31 December 1942 fears were expressed by Lieutenant General Joseph T. McNarney, Deputy Chief of Staff of the Army, in a memorandum to the Chief of the Air Staff:

> I note that practically without exception all service units being shipped overseas have fired only a familiarization course with the weapon with which they are armed. As I understand it, this familiarization course consists of only twenty rounds. It seems to me that an individual who has fired only twenty rounds would be more dangerous to himself and his fellow soldiers than to the enemy.[59]

★★★ 5 ★★★

# ORIGIN OF THE SERVICE GROUP

## JUNE 1942

### Antecedent Units

In the spring of 1942 a new organization was established by the Air Service Command to supplement all existing tactical units. This organization was to be known as a Service Group, and it was to man an installation called a Service Center. Its origin is of importance to this discussion, for it illustrates the continuity of the history of service organizations. It was developed not only from Air Depot Groups, but also from other units which had been known for many years in the peacetime army. These antecedent units had existed under the Air Service,[1] the Air Corps, and the GHQ Air Force.[2] The most important were Service Companies and Air Base Groups. Many of the subordinate units, which eventually joined Service Groups, had their origins in these older organizations.

The life history of a particular group, which survived the transitions and vicissitudes of the development of these units, would perhaps clarify the relationship of Service Companies (Squadrons) and Air Base Groups with the Service Groups of the war period. For example, a listing of the designations of the 3rd Service Group, as it came to be called, would fix in mind the sequence of events, for most of the changes in the logistical organizations of the air forces were reflected in the history of this group.

The designation of the original unit, which was later to become the 3rd Service Group, was the 57th Service Squadron.[3] This squadron was organized in 1922, shortly after World War I, at its home station, Selfridge Field, Mt. Clements, Michigan. All personnel were obtained from recruitment centers. In 1936 this unit became the 3rd Air Base Squadron, and as such it retained its identity until 1 September 1940, at which time the 3rd Air Base Group was formed. Under the command of Major Edgar T. Selzer, the group then consisted of four Air Corps squadrons and several attached units of other branches of the army. The guiding principle of all of these units was the policy of the GHQ Air Force that air bases should be "hotels for tactical units." The group was called the 3rd Air Base Group until the beginning of August 1942, when it was redesignated the 3rd Service Group at an overseas base.

The wartime disposition and utilization of the smaller units of Air Base Groups may also be illustrated by the story of this organization. From 19 August until 9 December 1941, several units of the 3rd Air Base Group took part in the extensive war-games already described, operating from temporary bases at Esler Field, Louisiana, and Spartanburg, South Carolina. Following the maneuvers, the two Materiel Squadrons of the group were separated, the 1st leaving for duty at Mitchel Field, New York, and the 2nd departing for the Port of Embarkation at San Francisco.

On 14 February 1942, the Headquarters Squadron of the 3rd Air Base Group departed by train from its home base at Selfridge Field, Michigan, for Florence Army Air Base in South Carolina. Units attached to the Squadron at the time of this move were the 726th Ordnance Company, Avn, (AB), and the 38th Signal Platoon (AB). Shortly afterwards the 1st Materiel Squadron rejoined the group from Mitchel Field, and all units began servicing the 52nd Pursuit Group, which was flying P-39's (Aircobras). Personnel of the Headquarters Squadron manned the base headquarters as an additional duty.

At this stage of the group's existence, enlisted men lived in pyramidal tents, and officers in smaller wall tents. The level of equipment and supplies was normal. Transportation and messing facilities were considered more than sufficient for the needs of the group. Soon after arrival at Florence, a light training program was started, which included lectures, calisthenics, athletics, and drill. Officers and non-coms, qualified to lead the group by special training or previous experience, were placed in charge of these activities. But training schedules were not permitted to interfere with normal working hours, for the group had work to do which had to come first.

On 28 April 1942 the Hq and Hq Squadron departed by motor convoy to the Wilmington Army Air Base in North Carolina. Here, in addition to the Ordnance Company and the Signal Platoon mentioned above, a Decontamination Detachment of the 1st Chemical Company was attached to the group, and the 1st Materiel Squadron rejoined the other units once more. At this base the group serviced two medium bomb group detachments on anti-submarine patrol over the Atlantic. The two

Air Corps units, however, were transferred almost immediately to Fort Dix, New Jersey, and upon assignment to the Eighth Air Force, they departed for the United Kingdom from the New York POE aboard the *Queen Elizabeth*. Not until after its arrival overseas was this Air Base Group redesignated a Service Group.

### The Service Test at Fort Dix

Unlike Air Depot Groups, Service Groups were never tested in army-wide maneuvers, though a few of the component units, as indicated above, did participate in the pre-war exercises in the South. Rather, the new organizations were shaped in a limited-scale test at Fort Dix, New Jersey, in May and June of 1942. This test was undertaken in an effort to cut the personnel and equipment of air force service organizations by at least 35 percent, in compliance with a request voiced by General Arnold early in the war.[4] It had become impossible at that point for the United States to transport its air force units in sufficient quantities, as they then existed. Consequently, it was directed that the air forces reorganize their tactical and logistical arrangements, and that unskilled basic soldiers and guards be eliminated from the T/O's of service units.[5] In accomplishing this change particular attention was to be paid to the proper employment of skilled personnel.

Plans for reorganization, in the process of formulation during the months of March and April, had a direct bearing on the Service Test at Fort Dix. On 24 March 1942 the Air Service Command was directed by Headquarters, Army Air Forces, to submit its recommendations for a revamped logistical organization in the light of several proposed changes. In April a teletype from the Washington headquarters to the Air Service Command explained the background of the service test, as follows:

> Plans currently under consideration provide for combat squadrons to be reduced to a strength of approximately 100 enlisted men by a transfer of all personnel and equipment, not peculiar to the type of aircraft with which equipped, to an Airdrome Squadron, which would be one of eight to twelve similar units operating dispersed airdromes in the vicinity of a "Service Center" serving two or three tactical groups. The "Service Center" would include approximately 2300 men, in addition to the 275 men in each of the Airdrome Squadrons, and would include units of other services required for third echelon maintenance and supply. . . .
>
> Your comments are requested as to whether or not consideration should be

given to utilizing one to four Air Depot Groups to operate the fourth echelon maintenance depot in the theater of operations, and utilize a "Service Group" or other group with equipment similar to that of the Air Base Group to accomplish second and third echelon maintenance. . . . It appears that Air Depot Groups should be retained for the fourth echelon maintenance for which they have been trained, that third echelon maintenance should be accomplished in the "Service Centers" by Materiel Squadron personnel and equipment, and that second echelon maintenance should be accomplished under the Air Service Command by personnel and equipment taken over from the combat squadrons.[6]

Not all of these ideas were adopted. Second echelon work, for example, was never the responsibility of the Air Service Command, and only a few Airdrome Squadrons were ever assigned to that command for training.* Instead, Airdrome Squadrons were trained by other commands for duties essentially similar to those of the ground echelons of combat units. They were used chiefly during leap-frog or island-hopping operations in the Pacific, or when it was desirable in stabilized areas to obtain rapid concentration of air power for a short period, without limiting the speed of concentration to the mobility of the combat squadrons' ground echelons.[7]

In general, however, the main outlines of the forthcoming organization of the air forces were sketched in advance. The Field Service Section of the Air Service Command believed that personnel in the T/O's for Air Depot Groups could be utilized for the operation of the more technical, or fourth echelon, maintenance depots with only a slight modification of buildings and equipment.[8] Certain duplications in Organizational Equipment Lists would have to be eliminated. Other items, such as accessory and instrument repair facilities, would have to be increased.

On 29 April 1942 the Air Service Command submitted a formal plan, complete with charts and tables, for "logistics in support of a typical air force in a Theater of Operations." [9] In this plan it was contemplated that each Service Group would include a Hq and Hq Squadron, eight Airdrome Squadrons, each capable of second echelon maintenance and supply for one combat squadron, and two Service Squadrons, each capable of third echelon work for one combat group. Fourth echelon depots would normally include a Repair and a Supply Squadron for each

* Most Airdrome Squadrons were trained by the domestic air forces, particularly the second. A few were trained by the Troop Carrier Command and the Air Service Command.

three tactical squadrons, and a Hq and Hq Squadron, Air Depot Group, for each three Repair and Supply Squadrons. It was understood that Repair Squadrons would include an extra 125 men each for engine overhaul.

Major General H. J. F. Miller explained in this communication that the demands of a primarily defensive war would be met, since active airdromes would be dispersed for protection and concealment. Moreover, personnel for both third and fourth echelon work would be decreased by 35 percent, as directed by General Arnold.

While these plans were being considered, the Air Service Command was ordered to "establish and service test without delay a so-called service center at the Fort Dix Flying Field for a period of approximately one month." [10] In conducting this service test, it was stated, there was to be no interference with the operations of the First Air Force and the Eastern Defense Command, which were cooperating in the exercise. Approval of the Commanding General of the New York Port of Embarkation, and arrangements with the Commanding Officer of Fort Dix for the establishment of a camp site at that location, had been obtained by Headquarters, Army Air Forces.

Lieutenant Colonel Paul E. Ruestow, Commanding Officer of the 91st Air Base Group, Mitchel Field, was assigned to the Air Service Command to take charge of the new organization, and other personnel of the Headquarters Squadron of that group were selected to accompany him. Tentative T/O's and T/E's for the Provisional Service Group were prescribed by the Director of Base Services; for the time being the group was to combine its new, tentative T/O's with older, official directives for Air Base Groups. After the experiment, comments and recommendations on manning, equipment, and the procedures of supply and maintenance, were to be submitted to the Director of Base Services.

Early in May 1942 the 3rd Materiel Squadron of the 91st Air Base Group was ordered to Fort Dix. This squadron was well qualified, for it had participated in the Louisiana-Carolina Maneuvers of 1941, having shared a hangar with the 4th Air Depot Group at Jackson, Mississippi. Part of the 6th Air Depot Group was transferred at the same time to provide personnel for the group's headquarters. Additional services, such as Quartermaster Truck Companies, were likewise made available to the 91st, as were transport airplanes to assist in the service of supply.

The following organizations, located at the eight stations indicated in

Table 7,[11] were to be supported by the new Service Center. These organizations were all within 125 miles of Fort Dix and were actively engaged in coastal defense.

TABLE 7

UNITS SERVICED BY THE 1ST PROVISIONAL SERVICE GROUP

| *Airport* | *Units Stationed* |
|---|---|
| Baltimore, Maryland | 59th Pursuit Sq. |
| Dover, Delaware | 112th Observation Sq. |
| Atlantic City, New Jersey | 104th Observation Sq. |
| Bendix, New Jersey | Interceptor Control Sq.; Hq & Hq Sq, 56th Pursuit Gp.; 62nd Pursuit Sq. |
| Newark, New Jersey | Detachment Interceptor Command |
| Philadelphia, Pennsylvania | Hq & Hq Sq, 33rd Pursuit Gp.; Interceptor Control Sq. |
| Fort Dix, New Jersey | 126th Observation Sq.; 119th Observation Sq.; 9th Observation Sq.; Hq & Hq Sq, 59th Observation Gp. |
| Millville, New Jersey | Class 4 airport to be ready for occupancy shortly. (An ideal spot to test Advance Detachment and Airdrome Sq.) |
| *Supplies to be received from:* | |
| Middletown, Pennsylvania | Air Depot |
| Fort Dix, then Mitchel Field, New York | Q.M. Depot |
| Mitchel Field, New York | Ord. Depot |

Meanwhile a conference was held in Washington, attended by Lieutenant Colonel Paul E. Ruestow, Lieutenant Colonel H. Paul Dellinger, and Lieutenant R. R. Scott, with Major General Carl Spaatz as a visitor. The purpose was to consider the service test of the Air Depot Group (as it was still called) and the new logistical setup for rendering second and third echelon maintenance and supply. As a result of the conference, additional recommendations were made to the Director of Base Services, through the Commanding General, Air Service Command, on 21 April 1942.[12] The most important was that the Headquarters Squadrons and Materiel Squadrons of Air Base Groups be transferred to the Air Service Command without delay. If the Materiel Squadrons alone were transferred, the Army Air Forces would be faced with the problem on the one hand of disbanding Headquarters Squadrons of Air Base Groups, and on the other of organizing completely new Headquarters Squadrons for Service Groups, a procedure difficult to justify.

In accomplishing this transfer, it was foreseen, certain pitfalls would have to be avoided. One difficulty undoubtedly would be that Headquarters Squadron personnel currently engaged in base activities would in many cases be considered indispensable to the bases. It was recommended that these individuals be assigned to Air Base Squadrons to avoid transfer to the Air Service Command.[13] To implement this policy, directives would state that actual duty assignments, as of a date prior to issuance of the orders, would govern the transfers of both officers and enlisted men. Since only 118 enlisted men were authorized for Air Base Squadrons, it would be necessary to increase the number to approximately 350 in the orders turning Air Base Groups over to the Air Service Command.[14]

At the beginning of the discussion differences of opinion had prevailed as to the necessity for the reassignment of functions. Colonel Ruestow, for example, believed that an Air Base Group, less its Air Base Squadron, could be utilized to perform third echelon work, and that only fourth echelon work should be handled by the Air Service Command.[15] Thus, he believed that Air Base Groups should be left under the numbered Air Forces. He based this opinion on the job his Materiel Squadron had done in the 1941 maneuvers. The 3rd Materiel Squadron with which this test was to be conducted, he pointed out, was the same unit he had commanded in the earlier maneuvers, and the equipment it would receive at Fort Dix was the orthodox amount already authorized for a Materiel Squadron by T/BA.

The officers from the Air Service Command, on the other hand, believed that third and fourth echelon work should be assigned to two separate types of organization, both under the Air Service Command, and that it would be necessary to create a new tactical service unit to supplement Air Depot Groups. The latter opinion prevailed, and means were sought to implement the new "service centers." (It should be pointed out, however, that at the end of the war third echelon units were given back to the air forces.)

Major General Carl Spaatz objected strenuously at the Washington conference to the use of the term "service centers." [16] Accordingly, other titles were suggested. Colonel Harold A. McGinnis proposed the established army term "advanced depot" as a means of conveying the purpose of the organization. Soon after the conference, General Spaatz was sent to England, but he continued his opposition to the term "service center." In a letter to General Arnold, dated 5 July 1942,[17] he reiterated his position: "As far as I am concerned, I would much prefer retaining the designation Mobile Air Depot Group for these installations. This constant changing of terms leads to too much confusion." To this General Arnold replied on 14 August 1942:

> The name Service Center which you object to seems to have been universally adopted by everyone, as the term is natural and logical for the functions which this groupment of service units perform. While in a sense this Service Center is an advanced depot, it is undesirable to call it a mobile air depot group for the reason that it is organized and equipped differently from the air depot group, contains personnel and equipment for performing third echelon maintenance and supply, while the air depot group, which is farther to the rear, contains personnel and equipment for fourth echelon maintenance and supply, including engine overhaul. It would be very confusing, both in equipping and training these organizations, to have them designated the same.[18]

On 14 May 1942 the service test at Fort Dix was begun. Unfortunately, in choosing the first location, the provisional group ignored the necessity for camouflage,[19] and soon had to be moved to a new location in the woods surrounding the base. Shortly afterwards, the runways of the Fort Dix Flying Field were extended to include the group's new bivouac area, and it was moved again, but in spite of these false starts, a successful and well-camouflaged Service Center was eventually set up.[20]

Altogether 42 officers and 853 enlisted men were assigned to the group.[21] There were also 192 "pieces of transportation" out of 375 authorized. Among those on hand, 90 percent were 2½ ton trucks. The tactical units which participated in the exercises were as follows:

6th Air Depot Group, Hq and Hq Squadron
3rd Materiel Squadron
26th Signal Platoon
630th Quartermaster (863rd) Maintenance Unit
154th Quartermaster Company
703rd Quartermaster Company
717th Ordnance Platoon
CWS Service Detachment (Provisional Group) [22]

Of the many activities connected with this service test, the busiest were the Mobile Repair Units. Four of these units serviced the eight dispersed airdromes listed in Table 7, and another unit was contemplated.[23] Originally, they were dispatched from Fort Dix to wherever they were needed, and upon completion of a job returned to their proper station. Thus, one Mobile Repair Unit went to Millville, New Jersey, as an advance party to prepare for the coming of a Pursuit Squadron of 25 P-40 E's from Philadelphia. It remained five days, and returned to Fort Dix. Another went to Dover to change three engines and was back the following day. Another was sent to Newark to retrieve a "nose up" A-20, and went home the same afternoon. Later it was decided to send each unit out, and let it stay at its temporary location until there was another call for its services. When this plan was introduced, Mobile Repair Units were to remain away from their base indefinitely, eliminating much unnecessary travel.

One of the questions raised at Fort Dix was the matter of duplication of equipment.[24] The Ordnance Company, for example, had an automotive shop truck which accomplished the same work as the shop of the Quartermaster. It had a welding shop truck which served the same needs as the Service Group's welding shop. Large Freuhoff trailers and elaborate instrument repair shops were regarded as superfluous for use at dispersed squadron airdromes. The Ordnance supply truck and the automotive wrecker with the fixed heavy tram rail were also called useless for their purposes by personnel of the group.[25] On the other hand, it was necessary to increase the number of bomb service trucks. Field drying shelters for parachutes were also tested at this time and added to the group's equipment.[26]

Setting up a propeller department presented many problems. On 16 May 1942 Colonel Dellinger reported that "if the group had to move within two weeks, the propeller balancing equipment would never be used and perhaps part of it, if not all, would be lost." [27] There were available, however, three instruments: a protractor, a vibration tester, and a balancing arbor, which could be used in combination to set and balance propeller blades, in most cases without removing the blades from the airplanes. The latter two had already been utilized successfully at both Mitchel Field and the Fort Dix Airport by two engineers from the Propeller Laboratory at the Materiel Center, Wright Field, Ohio, who had instructed enlisted men in their purpose and use. The kit was adopted as an advantageous substitute for the standard equipment, as it weighed 50 pounds rather than 4500.

The greatest annoyance for the engineering officers, however, was the lack of electric current.[28] The instrument repair trailer had a generator capable of a 40-amperage output, but more than 60 amperes were frequently required. The result was a continuous burning of fuses and low voltage, which in turn burned up the equipment. The machine shop trailer had the same problem. Technical supply also needed more power, as it had twice the amount of tentage authorized, and all of it illuminated.

Several organizations proposed for assignment to Service Groups were not tested in this exercise. For example, there was no opportunity to use Engineer Battalions (Avn), or Chemical Companies (Air Operations), or Medical Detachments. Major General H. J. F. Miller recommended to the Commanding General, Army Air Forces,[29] therefore, that an effort be made to streamline the first two of these units, and to obtain the opinions of the Air Surgeon with reference to Medical Detachments. He believed that a single detachment, assigned to Headquarters Squadron, would be sufficient for all units of the group.

In June, Colonel Dellinger reported that on the whole the Provisional Service Group was functioning satisfactorily. The only persistent complaint was the shortage of spares, a circumstance not attributable to the group. Inexperience undoubtedly had caused many troubles in the beginning, but this had been overcome as the new procedures became more familiar. Colonel Dellinger's conclusions were summarized as follows:

> This service group can perform the mission with which it is currently charged, and with the equipment now assigned. The system for maintaining combat units is sound and will probably only require reassignments of personnel as situations change. For instance this particular group would probably have to be strengthened in *sheet metal workers* if it were serving aircraft in the combat zone; more *cable splicers* would be necessary also. The T/O is adequate. The proposed T/O's for service groups call for 63 officers and 1091 enlisted men. The Provisional Service Group is satisfactory with 853 men and about 45 officers. The proposed T/O has enough additional men to allow for inexperience (the Provisional Service Group experience level is rather high) and for work in the combat zone not now anticipated.[30]

Until 14 November 1942 the 91st Service Group remained at Fort Dix as a model for the training of other Service Groups. [31] Cadres were sent from all over the country, as many as three 100-man cadres at a time, to observe the operations of the group and attend a series of lectures and demonstrations by group personnel. Each cadre stayed approximately two weeks. Upon return to proper station, the cadres instructed the other members of their groups, and in some cases set up "parent groups" for the training of other groups. The curriculum, as planned by Colonel Ruestow, is given in Table 8.[32]

TABLE 8

SCHEDULE FOR TRAINING OF CADRES AT FORT DIX

(Maximum of 3 cadres at one time for a period of 2 weeks)

I. ORIENTATION. For all personnel
  Plan for Supply and Maintenance Service to Air Forces
II. ORGANIZATION AND FUNCTIONS OF SERVICE GROUPS AND TOUR OF INSTALLATIONS. For all personnel
III. S-4 (SUPPLY) FUNCTIONS. For all officers
IV. EXECUTIVE AND S-1 (PERSONNEL) FUNCTIONS. For all officers
V. S-2 (INTELLIGENCE) AND S-3 (TRAINING AND OPERATIONS) FUNCTIONS. For all officers
VI. DUTIES OF ADJUTANT. For all officers
VII. CAMOUFLAGE. For all personnel (practical demonstrations)
VIII. 2ND ECHELON AIRCRAFT MAINTENANCE. For all officers and engineering enlisted men
IX. AIR CORPS SUPPLY. For all officers and all supply enlisted men
  Organizational Supply
  Table I and II
  Principal Crew Chief: 72 hours
  Squadron Engineering Set: 10 days
  Service Group Table I and II: 30 days
X. MORALE WELFARE: PHYSICAL FITNESS. For all officers
XI. FIELD SANITATION. For all personnel
XII. LAYING OUT SERVICE GROUP CAMP. For all officers (Circulation)
XIII. SIGNAL COMMUNICATIONS. For all officers
XIV. QUARTERMASTER BN. ORGANIZATION AND FUNCTIONS. For all officers
XV. ORDNANCE COMPANY ORGANIZATION AND FUNCTIONS. For all officers

The lectures [33] were taken down verbatim by three stenographers, and in the spring of 1942 the *Fort Dix Manual* was produced from the transcripts. This was the first Service Group Training Manual to be published, and as such gained a wide distribution in all four Air Service Area Commands. Another manual prepared on the basis of the Fort Dix Service Test was AAF Field Manual 1–195, *The Service Center*, dated 26 September 1942.

It would be difficult, therefore, to overestimate the contribution of the original Service Group to the organization and training of all tactical service units in the Army Air Forces. Moreover, the usefulness of this group was not ended when it left Fort Dix. It was destined for additional duty of the same type at the AAF Tactical Center in Orlando, Florida, and eventually it saw action at Saipan in the Marianas.

### Transfer of Air Base Groups to the Air Service Command

The reorganization of AAF service units as a result of the field test at Fort Dix involved several types of units. The process began on 5 May 1942 when the Air Adjutant General directed that personnel of Hq and Hq Squadrons and Materiel Squadrons of all Air Base Groups in the continental limits be transferred immediately to the Air Service Command.[34] Only those squadrons under orders for foreign duty, or those which had arrived at or were en route to Ports of Embarkation, and Materiel Squadrons assigned to the Ferrying Command were exempted.

Provision was made for the transfer of base commanders and base staff officers to Air Base Squadrons, to avoid reassignment to the Air Service Command.[35] The Commanding General, Air Service Command, was to see that the current maintenance and supply obligations of the reorganized units would continue to be met. He was further to request the redesignation of Hq and Hq Squadrons and Materiel Squadrons of Air Base Groups not later than 1 July 1942, and to initiate the activation of such additional service units as might be necessary in the future.

The process of transferring these organizations to the Air Service Command was by no means easy. Although the numerical designations of the units were retained to preserve whatever *esprit de corps* might have been built up during the lifetime of Air Base Groups, and although the transfers were relatively simple to accomplish on paper, the policy of Headquarters, Army Air Forces, was not clear in specific cases, and caused much confusion. It was directed, for example, that personnel turned over to the Air Service Command would remain in the positions they had formerly held, until the newly activated Service Groups were fairly well established.[36] Training of Service Groups was to begin at the stations where the men were then located. In some cases, however, unusual situations developed, for the new Service Groups were not ready to go into the field, and technical training had to be started at subdepots, which had previously had little to do with tactical organizations.

The question of precisely which personnel to transfer, and to whom, was to be decided by "mutual agreement to be reached between the station or training center and the commanding officer of the sub-depot or other representative of the Commanding General, Air Service Command." [37] Unfortunately, no provision was made for the possibility that "mutual agreement" might not immediately be reached, and the lower echelons could not obtain clarification when disputes were referred to Headquarters, Army Air Forces, for adjudication.

This was not a complete transfer of all Air Base Group personnel; rather, it was based on operational functions. But it was never defined by particular specialty serial numbers. For instance, the understanding of the Cochran Field Sub-depot in Georgia was that "transfer of personnel should include only those individuals connected with functions of supply and maintenance of aircraft, and such additional administrative personnel of Air Base Groups as deemed appropriate." [38] On the other hand, personnel at post headquarters, such as the post commanders and their staffs, technical inspectors, and operations personnel, would remain under their former command and ultimately be absorbed by Air Base Squadrons. Still, there was a wide area for disagreement and inter-command disputes.

The trouble was that base requirements varied in different parts of the country, and no machinery existed in 1942 to determine an equitable and exact manning table for each installation. Several false starts were made, but in the end the matter of requirements had to be left to the discretion of individual commanders.[39] At one point Air Base Squadrons were authorized more than 300 enlisted men, but on 27 May 1942 the Air Service Command ruled that they would have to be cut to 18 officers and 260 enlisted men.[40] On 10 June 1942 this policy was reversed, and it was stated that base commanders were not to be held to the figure shown in previous teletypes.[41] But no specific alternative was advanced, and the confusion persisted for some time.

The Air Service Command was reasonably certain as to the number needed for Service Groups, and the Combat Command was equally definite as to the requirements for the bases. The difficulty was that there was no impartial judge to settle disputed cases, and no directive to prevent the Combat Command from keeping the best men. The attitude of the Air Service Command was explained in a teletype from the Chief of the Field Service Section to the Commander of the Third Air Service Area Command in June:

While it is not the intention of this command to be arbitrary and unreasonable, it must, of necessity, devote its attention to its primary mission, and take prudent, reasonable, and energetic action to accomplish its mission. Consequently, unreasonable demands on the part of base commanders in the division of Headquarters and Headquarters Squadrons, and the transfer of personnel from Materiel Squadrons will not be agreed to by this command. . . .[42]

Jurisdictional disputes persisted, but the situation was permitted to remain *in statu quo.* On 13 June 1942 the Chief of the Field Service Section received from the Commander of the Third Air Service Area Command one of the most remarkable communications ever sent by a subordinate to a superior, describing in detail the transfer of Air Corps personnel in the course of this reorganization.

This headquarters being deluged and beseiged with communications and personal visits from sub-depot commanders, base commanders, control depots, technical commands, training commands, the Third Air Force, and units thereof, with reference to action to be taken in connection with taking over the Air Base Groups by the Air Service Command. . . . Teletype received from Commanding Officer San Antonio is quoted in part:

. . . Various station commanders feel that they do not have competent instructions and authority to proceed with arrangements for necessary transfers. . . .

Preliminary information received from Hendricks Sub-depot Commander, Sebring, Florida, indicates that the Air Service Command will obtain a total of less than 100 men and only 1 officer from the 56th Air Base Group located at that station. Only 21 of these men have had mechanical training. Reports received from other school activities indicate situations almost as bad, and many of the combat activities are asking for all or nearly all of the key men and most of the officers concerned. . . . In view of the instructions contained in a communication received from the Chief, Air Service Command, and various other communications received from your headquarters from time to time, which in effect direct that the various commands are to obtain or retain any and all personnel that they see fit, there seems to be nothing that can be done in the situation by this headquarters. . . . Training that was in progress is being interfered with, and future plans are being held up, combat group and smaller combat organizations are being moved from one part of the country to another, requests are being received for the transfer of individuals, and small cadres of men, and the war is going on, yet no decisive and energetic action is being taken by anyone to clarify this situation. . . . It appears that men could be obtained from Replacement Centers by the Air Service Command with almost as high experience level, with greater potentialities for future development, and with much less friction and discord than is the case, when using the present method of taking groups of men from the Combat Command that have already been stripped of the men with higher qualifications and inherent abilities. . . . Two logical solutions appear to present themselves:

1. To call the whole program off;
2. To take over the Bases together with the Base Groups *

Pending arrival at the final solution, request reply to questions presented in our teletypes . . . and any other instruction or information that will tend to clarify this situation.

THIRD AIR SERVICE AREA COMMAND [43]

The transfer was effected under these conditions, nonetheless, and everyone was forced to make the best of it. On 13 June 1942 the War Department directed redesignation of Headquarters Squadrons of Air Base Groups as Hq and Hq Squadrons, Service Group.[44] The same letter stated that Materiel Squadrons would be redesignated Service Squadrons. The corresponding units overseas were officially redesignated one month later,[45] although in a few instances the superseded designations were retained.[46]

The general reorganization of air force service units contemplated in June 1942, however, involved the transfer not only of Air Corps units, but also the units of other arms and services. Indeed, the T/O's of all Combat, Observation, Transport, Photographic, Air Base, and Air Depot Groups in the Army Air Forces were to be revised, effective 1 July 1942.[47] These changes involved the personnel and equipment of Chemical, Ordnance, Quartermaster, Signal, Finance, and Medical sections and/or units which composed or were attached to all existing air force organizations. Effective 22 June 1942, other arms and service units of the following types were transferred *en masse* (and with very little controversy) to the jurisdiction of the Air Service Command:

Quartermaster Company, Truck
Quartermaster Company, Light Maintenance
Quartermaster Platoon, Service Center
Ordnance Company, Air Base
Signal Platoon, Air Base
Chemical Platoon, Service Center
Chemical Company, Air Operations
Chemical Company, Air Bomb [48]

* This suggestion was made at other times and for other reasons, but it was never acted upon.

Henceforth, the Commanding General, Air Service Command, would establish Service Centers with both Air Corps and other arms and service units, and would accomplish the training of all units required for the operation of overseas Air Force Service Commands.[49] It became his further responsibility to allocate to the Commanding Generals of the air forces and commands whatever units were needed for tactical purposes, and to provide essential services which could not be provided by air base organizations and facilities. For allocating such units the following priorities were established:

a. Overseas task forces
b. 1st and 4th Air Forces
c. 2nd and 3rd Air Forces
d. Other activities [50]

## Plan for a Unified Logistical System

On 14 August 1942 AAF Regulation No. 65–1 was published to set forth the new system of "Supply and Maintenance of Army Air Force Units." This regulation was prepared by Colonel Paul E. Ruestow, Commanding Officer of the 1st Provisional Service Group, who had become Chief of the Logistics Planning Branch, Office of the Assistant Chief, Air Staff, Materiel, Maintenance, and Distribution. Although it was later slightly altered to meet changing conditions, its chief features remained in effect throughout World War II.[51]

Under AAF Regulation No. 65–1, and the T/O's of 1 July 1942, combat units were to be self-sustaining entities, performing first (the air) and second (the ground) echelon of maintenance and supply. Their responsibility included administration and housekeeping at dispersed squadron airdromes, except in the few cases where these functions were turned over to Airdrome Squadrons. Under the Air Service Command the new Service Group was to be responsible for third echelon work, which would consist of repair and salvage beyond the capabilities of second echelon, including the operation of Mobile Repair Units, but not the establishment of Distribution Points at combat unit airdromes. Fourth echelon maintenance and supply was to be provided by the reorganized Air Depot Group, which was to be "in the rear of," or supporting, one or more Service Groups.

The organizational plans of the Army Air Forces, as outlined in this regulation, were predicated on certain assumptions and certain observations, which were made both prior to our entry into the war and during the first few months of our participation.[52] Most important of all was the fact that the Germans and the Japanese had obtained superiority in the air. Destruction of American and allied aircraft on the ground had led to a concept of operations from individual squadron airdromes, to secure dispersion and concealment. Air fields were available at that time; combat units were scarce. The tendency was to think in terms of dispersed squadron airdromes and wing (or multi-group) headquarters. Not until after we had achieved air superiority of our own was it possible to locate an entire combat group, or perhaps two combat groups, at a single air base within range of the enemy. When airplanes were furnished to Theater Commanders in numbers sufficient to permit, or even to require, the concentration of combat units, less stress was placed on dispersion.

Factors peculiar to the nature of air operations had lent themselves to the creation of a logistical organization different from that of the Ground Forces. (See Fig. IV.) It was impossible, for example, to draw a line on a map to separate the combat zone from the communications zone of an overseas Air Force Service Command. Strategic and tactical air forces were necessarily intermingled, with air units adjacent to ground units they did not directly support. The range of the air arm and the geographical characteristics of the theaters resulted in the superimposition of combat and communications zone installations one upon the other. This condition was to be particularly noticeable in England, where some units of the offensive weapon (the Strategic Air Force) were well within the service area.

Air Force Service Commands had to be flexible, and to utilize both ground and air force facilities, or to be self-sufficient, as the case might require.[53] Moreover, the tempo of air operations had to be geared to the speed of new and constantly changing airplanes. The weight and volume of fuel, bombs, and ammunition were enormous, and the activity of the ground crews was greatly stepped up just before and just after each combat mission. Stationary air bases, like naval bases, had to be prepared for the sudden and sometimes unexpected arrival of damaged ships and new personnel, requiring hospitalization, housing, and food.

The plan set forth in AAF Regulation No. 65-1 was to be a guide for all service units of the Army Air Forces in this country and overseas. First and second echelon work was to be done by the numbered air

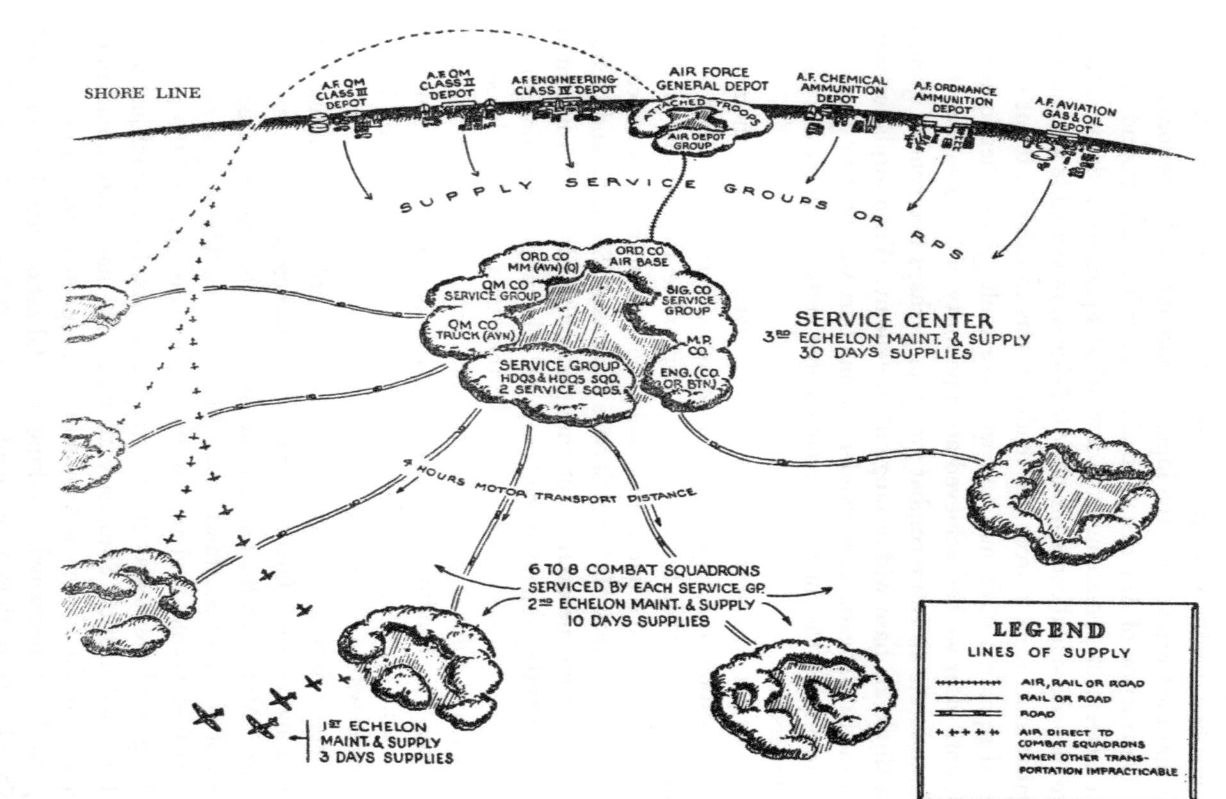


FIGURE IV

AIR FORCE SUPPLY IN SERVICE CENTER AREA

forces, third and fourth by the Air Service Command or its equivalent. The main outlines of this system were to be applied to all U.S. Air Forces everywhere in the world. Only when local conditions were prohibitive were variations to be attempted.

# 6

## OTHER ARMS AND SERVICES

### Unit Training

Unit training was designed to prepare subordinate units for service as component parts of larger organizations. Ordinarily it was conducted before the units were assembled into groups for final phase or combined training. In the case of arms and services with the Army Air Forces (ASWAAF) units, preliminary training was often conducted at dispersed bases under the other branches of the army. For some of these units the Air Service Command assumed no direct responsibility until after the formation of complete groups, but it was always interested in their welfare, for the units of the other arms and services were often the backbone of AAF service organizations.

The majority of ASWAAF units were organized to serve with regular Air Depot and Service Groups, but some existed separately to perform unique functions, and were assigned "one per army," or "one per air force," or "as required." Engineer and Military Police units, for instance, were almost always independent; only on rare occasions were they attached to commands as small as the headquarters of tactical groups. If such units were at any time attached to these groups, however, they have been included in this discussion. The services which they performed were important; in fact, the development of AAF service organizations cannot be understood without a comprehension of the duties of the other branches. The AAF may have wished for independence, but in the war years it was very much a part of the army, and received substantial help from the other arms and services, especially the following:

| | |
|---|---|
| Chemical Warfare | Military Police |
| Engineers | Ordnance |
| Finance | Quartermaster |
| Medical | Signal |

CHEMICAL WARFARE

The mission of the Chemical Warfare Service (CWS) with the Army Air Forces was threefold: the protection of the air forces against chemical attack, the giving of advice and recommendations for the offensive use of chemical warfare, and the supplying of all chemical equipment, munitions, and agents.[1] To accomplish this mission chemical officers were assigned to staff positions in air force headquarters and in headquarters of subordinate commands down to the level of wing. Below that echelon officers and non-commissioned officers engaged in this work were referred to as "unit gas officers," and their duties were considered as part-time.[2] Within the Air Service Command CWS officers were responsible for administration, technical intelligence, operations, training, and supply.[3] There were several tactical units to be trained, usually as regular features of groups.

On 22 June 1942 all Chemical Companies, Air Operations, together with Chemical Companies, Air Bomb, and Chemical Platoons, Service Center, were transferred from the numbered air forces to the Air Service Command.[4] The latter two types of units were disbanded,[5] and their personnel and equipment were absorbed by Chemical Companies, Air Operations. On 1 September Chemical Depot Companies (Avn) and Chemical Maintenance Companies (Avn) were also assigned to the Air Service Command, as was further responsibility for their training.[6] The Air Service Command was thus responsible for the training of three primary units:

Chemical Company, Air Operations (T/O 3–457)
Chemical Depot Company (Avn) (T/O 3–418)
Chemical Maintenance Company (Avn) (T/O 3–147)

The first of these units was a filling company, which filled airplane spray tanks with mustard gas or other chemical agents and furnished them to the airplanes. After the spraying assignments, this unit decontaminated the tanks and stored them for reuse.[7] In 1943 one Air Operations Company was authorized, when required, for each combat group; it was composed of four platoons, one for each combat squadron.[8] The Chemical Depot Company (Avn) handled Class V supplies, consisting of chemical ammunition and pyrotechnics, and operated Chemical Ammunition Depots in which toxic gases, smokes, and incendiaries were stored. The Chemical Maintenance Company (Avn), was responsible for third and fourth echelon repairs on chemical equipment and acted as a salvage and repair unit.[9]

To make the most efficient use of available personnel and equipment, it was decided that chemical companies should be concentrated at two primary training centers: one at Merced, California (later moved to Reno, Nevada, then to Fresno, California); the other at Herbert Smart Airport, Macon, Georgia. In addition, a training center for colored chemical troops was set up at Columbia Army Air Base, South Carolina.[10] This concentration of troops was considered to be a great improvement. The experience level of branch officers was low, however, and the problem of procuring training equipment was only partially solved. The first units were sent overseas from the new training centers in May 1943.[11]

Special arrangements were made to facilitate the program. An agreement was made by Headquarters, AAF, with the Chief of the Chemical Warfare Service whereby newly activated ASC units would receive initial mobilization training at the CWS Unit Training Center, Camp Sibert, Alabama. Seven Air Operations Companies were ordered from Herbert Smart Airport to Camp Sibert on 21 January 1943 for fifteen weeks' training. This arrangement proved highly satisfactory, and was continued until the end of 1943. In all, twenty-three Chemical Companies, Air Operations, three Chemical Depot Companies (Avn), and one Chemical Maintenance Company (Avn) received mobilization training at Camp Sibert.[12]

Another inter-command arrangement was that entered into by the Warner Robins Air Depot with the Third Air Force to provide Air Operations Companies with actual practice in the filling and handling of smoke tanks in connection with bombardment operations.[13] A program of combined training with the medium and light bombardment training units of the Third Air Force was arranged for the mutual benefit of the two commands. Similar arrangements were made by the Sacramento area with AAF activities on the West Coast and with the Desert Training Center.

Altogether 117 CWS units were trained by the Air Service Command, of which 46 were sent overseas: 33 Chemical Companies, Air Operations; 9 Chemical Depot Companies (Avn); and 4 Chemical Maintenance Companies (Avn). None of the CWS units shipped to the Ports of Embarkation failed to pass their final inspections.[14]

ENGINEERS

The basic mission of combat Engineers—ground or aviation—was "to assist the movement of our own forces and hinder the movement of the enemy." [15] At the same time they were expected to increase the combat effectiveness of American or Allied troops by construction measures for the improvement of health, comfort, and safety. "Movement" for the air forces meant new forward bases from which to extend their operations. The main job of Aviation Engineers, therefore, was to construct, maintain, defend while constructing, conceal, and if necessary destroy these bases.[16] At Salerno, Italy, for example, a fighter landing strip was ready by D plus 2, and was of inestimable value in the air support of the ground troops.[17]

The basic construction unit for this type of work was the Aviation Engineer Battalion, a large working and fighting force of approximately 800 officers and men, with 175 pieces of heavy construction equipment.[18] There were also other units for camouflage duties, mapping and topographical work, and airborne operations, all of which served directly under air force headquarters. In rear areas, general service Engineer organizations took over from the combat Engineers, who then moved forward. General service units, however, were not trained by the Army Air Forces. Altogether, the principal Corps of Engineer (CE) units which were trained by the Army Air Forces included:

Engineer Aviation Regiment
Engineer Aviation Battalion (T/O 5–415)
Engineer Aviation Company (T/O 5–417)
Engineer Air Force Headquarters Company (T/O 5–800–2)
Engineer Aviation Topographic Company (Old T/O 5–447) (New T/O 5–400)
Engineer Aviation Camouflage Battalion (T/O 5–465)
Engineer Airborne Aviation Battalion (T/O 5–455) [19]

The training of Engineer units in services peculiar to the Army Air Forces was almost non-existent until the spring of 1941. At that time the First Air Force was assigned the control of the 21st Engineer Regiment, and shortly thereafter began the training of Engineer Aviation Battalions at Westover Field, Massachusetts.[20] Training Battalions were subsequently activated by the other numbered air forces. The earlier battalions were fortunate in containing selected enlisted personnel who had received eight to twelve weeks of excellent basic military training at the Engineer Replacement Training Centers at Fort Belvoir, Virginia, and Fort Leonard Wood, Missouri.[21]

In 1942 the 21st Engineer Regiment at Westover Field developed into an Engineer Aviation Unit Training Center, which concentrated on specialties peculiar to the Army Air Forces. The other training battalions followed suit, but were soon unable to meet the growing demand for Engineer personnel, and fillers were requested from the Infantry and other branch training centers, and directly from Reception Centers. Basic military training was then added to the functions of the Engineer Aviation Unit Training Centers.[22] In November 1942, when basic training of other arms and services became temporarily the function of the AAF Technical Training Command, all Engineer basics were sent to Jefferson Barracks, Missouri, where they were trained under the technical supervision of the Air Engineer.[23]

The number of Aviation Unit Training Centers was reduced in 1943 to two: the Fourth Air Force Center at Geiger Field, Washington, and the Third Air Force Center at MacDill Field, Florida; these training centers specialized in the training of white and colored personnel, respectively.[24] Ordinarily, the training period for committed units was at least twenty-four weeks: thirteen weeks of basic military training and eleven weeks of unit training. The majority of Engineer units were trained at the two Air Force Unit Training Centers and at such specialized installations as AAFTAC and the Desert Training Center. The timing of this program was important, for Aviation Engineers were supposed to complete their construction work in the field before the arrival of any other units.[25]

The training of Airborne Battalions began in the fall of 1942 after considerable experimentation.[26] This program was essentially the same as that of the regular battalions, although a slightly longer period of time was required. The Airborne Battalion was the one Engineer unit requiring extensive combined training; the Troop Carrier Command was given this responsibility. The program ranged from six to eight weeks, with the first portion devoted to combined training, and the second to maneuvers with the Airborne Command of the Army Ground Forces.[27] The success of these endeavors was illustrated in the operations against Lae and Wewak in the South Pacific, during which a fighter field was constructed forty miles from Lae, at a site inaccessible except by air.[28]

Another Engineer activity important to the Army Air Forces was the supervision of all camouflage training. Battalions were trained for this purpose at the Camouflage Schools at Walterboro, South Carolina (later at Dale Mabry Field, Florida) and at Hamilton (later March) Field, California. These battalions were stationed at different times under the various commands and air forces in continental United States; and companies and detachments thereof were dispatched to dispersed air fields to supervise instruction and perform demonstrations. In the latter part of 1944 all formal camouflage training was concentrated in the school at March Field, although the technical training of officers was continued in a two weeks' course at Fort Belvoir, Virginia. Two training aids, the Third Air Force puppet show "Snafu" and the Fourth Air Force stage production "You Bet Your Life" traveled extensively and were considered successful both as instructional media and as entertainment.[29] In the Air Service Command the Engineer Section helped in the publication of a camouflage training manual that was used by many tactical service units.[30]

## FINANCE

The need for Finance personnel was supplied by the inclusion of Finance Sections in the T/O's of Hq and Hq Squadrons. In Air Depot Groups prior to January 1944 Finance Sections consisted of one officer and five enlisted men. This number proved insufficient, for no allowance was made for absences caused by sickness or furloughs. On 20 January 1944, the T/O was increased to two officers and fifteen enlisted men, and the Air Service Command was requested to furnish the extras needed to fill the groups already overseas.[31] The Finance Sections of the old type Service Groups consisted of two officers (one Major and one 1st Lieutenant) and fifteen enlisted men, a T/O which proved sufficient and was never changed. New Type Service Groups required two officers (one Captain and one 1st Lieutenant) and ten enlisted men.

The mission of the Finance Section of a tactical group was to provide prompt Finance service for the units of the group, other units in the Service or Air Depot Group Area, and such other organizations as higher authority might direct. This service included primarily the disbursement of public funds to military and civilian personnel for services rendered, the supervision and maintenance of fiscal records for payments made and balances of funds, advice on technical questions pertaining to the Finance Department, the preparation and verification of all classes of payment vouchers, and, in some theaters, responsibility for the audit of military property accounts as ordered by the theater commanders.[32]

While the payment of troops was a command function, it was the duty of the disbursing officer to assist in maintaining morale by promptness and efficiency in the payment of all accounts, and by proper coordination with the personnel officers responsible for pay and mileage forms. The duties of the disbursing office in the matter of property auditing were exacting and difficult in this country, but were greatly simplified in the theaters.[33] In both cases the disbursing officer of an air force service center had functions equivalent to those of a division Finance officer in the ground forces, and was governed by the same basic regulations, AR 35-5 and AR 170-10. Only in one respect was his function peculiar, and that was in the payment of flying personnel.[34]

In the early days of the service unit program, no trained Finance personnel were available within the Air Service Command, and a request was made of the Army Service Forces Training Center, Fort Benjamin Harrison, Indiana, for personnel in the form of preactivated units. These preactivated units consisted of the full T/O complement: two commissioned graduates of the Finance OCS at Fort Benjamin Harrison, and fifteen enlisted graduates of either the basic or the advanced technical school at the same station. The men were well trained, and it is of interest to note that Finance personnel had a higher average AGCT (intelligence) score than that of any other branch in the Army Air Forces.[35]

The greatest deficiency of Finance Sections overseas was a lack of accountants and bookkeepers to handle the groups' functions as accountable disbursing offices.[36] It was directed in June 1943 that at least two men from each group be thoroughly trained in this important work.[37] Since the use of dummy accounts and textbooks was not considered enough, the men were to be placed in near-by disbursing offices whenever possible for more responsible work. If the students were not to be allowed to keep the official books, they were to use the same vouchers and other accounting papers as the regular bookkeepers, and at the end of the day to compare their books with the official books for corrections.[38]

During the combined training phase, just before overseas movement, it was the responsibility of group Finance officers to provide for the requisition, receipt, utilization and/or storage of all office equipment, blank forms, regulations, and directives. Authority was obtained by the Air Service Command for each section to take with it a six months' supply of Finance and Treasury Department forms necessary for the immediate establishment of Finance offices at overseas locations. This action was taken because of the numerous complaints from the theaters that Finance officers were unable to operate because of the shortage of supplies.[39]

MEDICAL

The functions of Medical Department personnel in air service organizations included the usual Medical services (complicated by the industrial nature of these groups), and also Medical supply, and evacuation.[40] The general program of Medical training was broken down into five categories:

Training for Non-Medical personnel
Training for Medical personnel
Training for Non-Medical units
Training for Medical units
Training in Industrial Hygiene [41]

Training for Non-Medical personnel, as elsewhere in the Army Air Forces, dealt primarily with first aid, sanitation and hygiene, adjustment problems, and altitude indoctrination. This training was accomplished by the Air Service Command in the usual way except in two particulars; these were a special, comprehensive program of instruction on malaria control and malaria discipline, which was begun in January 1944 by all area commands to prevent further high incidence of the disease in overseas theaters, and a series of short courses on altitude indoctrination to give flying personnel assigned to administrative duties the benefit of the latest instruction given to combat crews.

The training of Medical Department personnel was conducted early in the war by the commandant of each AAF hospital. Most of the incoming doctors and dentists were fresh from civilian life and had very little knowledge of military procedure. Few had been to the Medical Field Service School at Carlisle Barracks, Pennsylvania. Enlisted personnel had even less background for their jobs. In the early days it was not unusual for ASC hospitals to receive cadres of varying sizes, consisting of enlisted men from all walks of life, with all levels of intelligence, and all degrees of training. Many of these men had had no more than two to four weeks' instruction.

Allotments to the various ASF Medical technicians' training schools were utilized to the fullest extent, but on-the-job training was all that could be attempted in many AAF hospitals, because there was a shortage of personnel in these establishments, and in some instances the quotas were not large enough to fill the needs of the particular commands. Moreover, in August 1943 the use of ASF Medical schools was discontinued throughout the Army Air Forces, and thereafter all commands had to use on-the-job training. This system was a help to a few hospitals, but experience soon demonstrated that personnel so trained were of little value in other hospitals without retraining, even though they were properly classified.

The training of Non-Medical units was largely a matter of individual indoctrination and "processing" in preparation for overseas service.[42] There was little unit Medical training as such. Responsibility for processing devolved upon the Medical Section of each group; these sections were trained as separate Medical Department units. Thus, the training of Non-Medical units merged with the training of Medical units of which, in addition to group sections, there were three types: Medical Supply Platoons (Avn), Medical Dispensaries (Avn), and Medical Squadrons (Air Evacuation).

Medical Supply Platoons (Avn) were trained in 1942 and 1943 by the Medical Training Section at Warner Robins, Georgia.[43] The original units were little more than aggregations of untrained individuals with many classifications brought together and officially designated as Medical Department units. The personnel had few, if any, qualifications for their jobs. As time went on, training and classification at this location became more efficient, but this very efficiency had a drawback. The training period was six months, the same as that of the other elements of the groups. These small units, however, had usually received what little unit training they needed within the first two months. After that they had no new responsibilities, and their internal organization disintegrated, as did morale and efficiency. Yet it was not until late in the fall of 1944 that authority was granted to reduce the length of the training period from six months to two. Medical Dispensaries (Avn) were also trained for a short period at Warner Robins, while Medical Squadrons (Air Evacuation) were trained at Bowman Field, Kentucky, in conjunction with the Troop Carrier Command.

The ASC Medical Section had its hardest job in the training of group Medical Sections for squadron and group aid stations.[44] These sections were set up for "general practice," and bore something of the same relationship to the other Medical Department units that a family doctor bears to a hospital specialist. In 1942 the Surgeon at Warner Robins was given authority to set up a centralized Medical Training School to accommodate the personnel of the several groups then being trained at that depot. Later, Medical officers reporting for duty with the Air Service Command were assigned to this school, and thence transferred to tactical units of the command or to fixed installations. Enlisted personnel were received from three sources: directly from Headquarters, AAF; from ASC station hospitals; and from ASF schools. Among its training facilities the school had a large, well-constructed sanitary area, which subsequently became a model for all other sanitary areas in the command and in most of the Army Air Forces. It also had a number of CCC buildings and a few newer structures, which were used as classrooms, offices, and barracks. It was manned by personnel of the station hospital.[45]

This installation was satisfactory so far as facilities were concerned, but proper arrangements were never made for the assignment of personnel. Doctors arrived at an unpredictable rate: sometimes one a day, sometimes fifty. Thus, previously planned Medical Sections had to be broken up, and the general confusion led to maladministration. In December 1942 it was discovered that there were eighty Medical officers at Warner Robins who had been through the school once, and had practically completed the same course again. In this group there were officers who had been to the Medical Field Service Training School at Carlisle Barracks as well as to the Warner Robins school, and then through some mistake had been sent back to take the same course. The morale of these doctors was hardly fit to be described, and this situation, along with several similar instances, did much to destroy confidence in the good work of the school. Because of this failure, the surgeons at other control depots began to distrust the products of the school and to set up their own training programs.

To avoid the duplication and waste of having separate schools in each area, the ASC Surgeon set up what he hoped would be "the Carlisle of the Army Air Forces." On 10 November 1943, after more than 1,000 officers and 6,000 enlisted men had been trained and sent overseas, the AAF Medical Service Training School (subsequently the 25th AAF Base Unit) was officially established at Warner Robins, and directed to undertake the training of all Medical Department officers and the preparation of enlisted cadres for all tactical units. For administrative purposes the school was placed under the Commanding General, Air Service Command.[46] In addition, the ASC Medical Training Detachment No. 1 (later the 4520th AAF Base Unit) was established to assist in the training of Medical Sections, Medical Supply Platoons (Avn), and Medical Dispensaries (Avn) at a single centralized location.

One of the cardinal principles of these schools was that every man in the Medical Department, regardless of his classification, should learn to perform emergency medical treatment, to administer plasma, and to give hypodermic injections.[47] To fill the need for instructional material, a "Medical Officer's Guide" was written and distributed by the Air Service Command to all Medical Department personnel. This pamphlet was subsequently adopted as a standard text by the School of Air Evacuation at Bowman Field, Kentucky, which was engaged in the training of Medical Squadrons (Air Evacuation). It later became T.O. 30–1–5.[48]

In all of this training, it was imperative to achieve the same level of preventive industrial hygiene in service centers and air depots overseas as that already attained in the zone of the interior. The first attempt at meeting this challenge was undertaken at Oklahoma City, where a technique was evolved which was subsequently reproduced elsewhere.[49] This plan called for ten hours' training for Medical Department personnel and four hours' training for Non-Medical personnel; it consisted simply of tours through the supply and maintenance installations of the depot to point out the dangers of certain selected operations and the applicable preventive measures. To standardize this program, the Industrial Hygiene Branch, Office of the Surgeon, ASC, published several booklets.

### MILITARY POLICE

There were but two Military Police (MP) units in which the Army Air Forces was interested: Guard Squadrons and Military Police Companies (Avn). The former were trained to perform interior guard, traffic control, riot control and vice control at continental installations; the latter, to perform the same functions overseas. Originally, these units received their training at air force and command stations all over the country, with a minimum of central control. The personnel were trained by the Army Service Forces and also by the Army Air Forces. Occasionally they were transferred from the Air Corps to the Corps of Military Police.[50] Sometimes they were trained as tactical organizations, sometimes as parts of station complements.

Officers were furnished to the Air Service Command by the Army Service Forces under War Department policies for the assignment of Branch Immaterial personnel. Until the full quotas were acquired, however, Headquarters, AAF, was authorized to detail both Air Corps and Branch Immaterial officers to these organizations. Enlisted men were supplied originally in the form of cadres from the aviation guard units of the Army Service Forces, with fillers requisitioned from The Adjutant General in the usual manner.[51] Later they were transferred to Guard Squadrons from Air Corps Basic Training Centers, and to Military Police Companies (Avn) from the Basic Training Centers of the Technical Training Command.[52] Upon activation of Guard Squadrons or Military Police Companies at Air Corps stations the overstrength of fifty enlisted men authorized for Base Headquarters and Air Base Squadrons or for Hq and Hq Squadrons of Service Groups was automatically canceled.[53]

In May 1943 a Military Police Training Center (Avn) was established at Camp Ripley, Minnesota, under the administrative supervision of the Air Service Command. In January 1944 this center was moved to Camp Barkeley, Texas, where it functioned as part of the San Antonio ASC training structure and trained both Guard Squadrons and Military Police Companies (Avn).[54] No sooner was it moved, however, than Camp Barkeley was found to be unsuitable, and on 21 February 1944 it was transferred to Barksdale Field, Louisiana. Responsibility for Military Police training was then transferred from the Air Service Command to the Third Air Force.[55] Thereafter all Guard Squadrons and Military Police Companies (Avn) in the Army Air Forces were trained at Barksdale Field under the administrative supervision of the Third Air Force and the technical direction of the Provost Marshal General.

The organization of MP units was similar to that of civilian police forces. The chief activities were desk and records, service, investigations, traffic and gate control, dismounted patrol, and motorized patrol.

Personnel were formed into cadres at the Military Police Training Center (Avn), and subsequently transferred to the various continental stations, where they formed nuclei for new guard units.

In training these units, emphasis was placed on basic military training, especially physical development, marksmanship, close order drill, and defensive combat. Instruction in military law and criminal investigation was also included for both officers and enlisted men.[56] Ordinarily all personnel received a minimum of six hundred hours' training over a period of forty weeks. Whenever possible, stress was placed on on-the-job work, which involved the direction of local traffic and the patrolling of near-by towns.

### ORDNANCE

The mission of Aviation Ordnance was the supply and maintenance of armament, ammunition, and automotive vehicles.[57] Armament in the Army Air Forces included aircraft weapons, anti-aircraft and anti-tank weapons, and small arms. The first category, aircraft weapons, was the most important; with these, the air force accomplished its basic mission. Anti-aircraft weapons were included whenever Coast Artillery troops or anti-aircraft regiments were assigned to the air forces; anti-tank weapons were part of the equipment of Aviation Engineer Battalions. The third group, small arms, was present in all AAF organizations.

Ammunition required by the Army Air Forces fell into two categories.[58] First, that peculiar to the Army Air Forces—the munitions with which the missions were accomplished, notably aircraft bombs and accessories (fuses, fins, arming wires, and so on). The second class was common to Army Ground Forces, Army Service Forces, the Navy, and the Army Air Forces, as, for example, ammunition of .30, .45, and .50 calibre.

Air Force automotive vehicles included three general classes.[59] First were special purpose vehicles peculiar to the Army Air Forces, such as bomb service trucks, bomb lift trucks, and bomb trailers. Second were special purpose vehicles not peculiar to the Army Air Forces, such as machine shop trucks, automotive and small arm repair trucks, and so forth. The third group was common to all armed forces: cargo vehicles, personnel carriers, and "jeeps." Supply and maintenance was the responsibility of the Ordnance Department; operation of the vehicles was not. The only vehicles actually operated by the Ordnance Department were those assigned to Ordnance units. Motor pools and truck companies were operated by the Air Quartermaster, and by other agencies of the Army Air Forces. In this respect the weapons situation was analogous; although the Ordnance Department provided and maintained the weapons, it did not furnish the personnel to fire them.

On 22 June 1942, Ordnance Companies, Air Base, were transferred from the air forces to the Air Service Command, along with the other component units of Air Base Groups,[60] as part of the plan to make one command responsible for all third and fourth echelon maintenance and supply. Data regarding these companies were turned over to the Air Service Command by the air forces in very poor condition; in fact, all that could be learned was obtained from the latest manning charts on file in Washington. These proved to be faulty in many respects, but eventually correct data were obtained. During the succeeding months, according to the ASC Ordnance Officer,

> Shipments of Air Base Companies continued, and it is regretted to say that many departed with little or no training. In fact, they were lucky to have four (4) officers and a full complement of enlisted men, and in those days MOS numbers didn't matter—it was simply counting noses. In order to provide four (4) officers for each company, it was necessary at times to reduce some of the remaining companies to no officers at all, leaving the company in charge of the 1st. Sgt.[61]

In August, Ordnance Medium Maintenance Companies (Avn) (Q) were also transferred to the Air Service Command.[62] As the "(Q)" indicates, these companies had been under the Quartermaster General before the transaction took place. The change was occasioned by the decision to place all third and fourth echelon vehicle maintenance under the Ordnance Department, leaving only the operation of the vehicles and first and second echelon work to the Quartermaster (a division of functions similar to that obtaining in the Air Forces). Then in October, fifteen of the three following types of companies were transferred: Ordnance Ammunition Companies (Avn), Ordnance Depot Companies (Avn), and Ordnance Maintenance Companies, Air Force. The last at that time was known as the Ordnance Medium Maintenance Company, but the similarity of its name to that of the Q Company caused much confusion, and the designation of the larger company was changed. Finally, in the spring of 1943 another reorganization was effected, and the Ordnance complement of each Service Group was changed from an Air Base Company, two Q Companies, and a Headquarters Section to two Supply and Maintenance Companies.[63] Although this reorganization had been caused by a desire for economies in manpower, the net result was the same number of Ordnance officers and the same number of Ordnance enlisted men in each group, but differently organized.

Altogether there were six Ordnance service organizations and three Ordnance headquarters sections in the Army Air Forces. During the course of the war, a total of 445 Ordnance units with 1,410 officers, 69 warrant officers, and 28,419 enlisted men was shipped overseas.[64] The units of primary importance were:

Ordnance Ammunition Company (T/O 9–17)
Ordnance Depot Company (T/O 9–57)
Ordnance Maintenance Company, Air Force (T/O 9–257)
Ordnance Supply and Maintenance Company (Avn) (T/O 9–417)
Ordnance Company (Avn) (Air Base) (T/O 9–167)
Ordnance Medium Maintenance Company (Avn) (Q)

In training these units, many of the usual problems arose,[65] especially the following:

(a) Scattering of small units over a wide geographical area;
(b) Lack of commissioned personnel other than recent and relatively inexperienced graduates of the OCS at Aberdeen Proving Ground, Aberdeen, Maryland;
(c) Insufficient capacity of Ordnance Replacement Training Centers to train cadres for newly activated air service units;
(d) Insufficiency of SOS school quotas prior to 1 December 1942;
(e) The necessity for using basics rather than trained specialists as fillers;
(f) The unsatisfactory quality of the classification tests at the Reception Centers.[66]

In view of these conditions, the Air Ordnance Officer felt that the wisdom of utilizing Ordnance Department facilities was incontestable. Accordingly, Ordnance units serving with the Army Air Forces were sent whenever possible to the Ordnance Unit Training Centers (OUTC's) at Flora, Mississippi, and Santa Anita, California. To avoid misunderstandings careful agreements were entered into by the two commands. Although the Air Service Command was to be responsible for the filling of the units, the Army Service Forces was to exercise command while the units remained at the Ordnance Unit Training Centers. Only if they remained at the OUTC's for 13 weeks was the Army Service

Forces to assume responsibility, but even in that event inspections by AAF representatives were to be authorized. If a unit received its Movement Orders while still at the OUTC, the Army Service Forces would be responsible for all its final preparations for overseas movement.[67]

### QUARTERMASTER

The general function of Air Quartermasters was to furnish the necessary personnel and facilities for the proper Quartermaster supply of all air force troops in the theater of operations.[68] Quartermaster supply in 1943 was broken down into seventy-four categories, which for greater convenience were often assembled into the five main War Department classes: Class I, subsistence; Class II, T/BA and T/E issue of clothing, equipment and supplies; Class III, fuels, aviation and automotive; Class IV, articles not on the tables, including construction and camouflage material; and Class V, ammunition, chemicals, and pyrotechnics.

The manner in which Class I supplies (rations) were distributed was roughly the same in every theater.[69] Seventy-two hours in advance of the need, each squadron telegraphed or telephoned its ration strength to the Quartermaster at the appropriate service center. These returns were consolidated there and sent to the Subsistence Company, or the Class I depot, which was often under the Services of Supply (Army Service Forces). The day prior to use, the rations were issued at the depot in bulk and forwarded by truck to the service centers, where they were broken down and shipped by truck to the using organizations at the operating airdromes. Class III supplies were procured in the same manner, the only exceptions being that requirements were based on quantities consumed the previous day, and the requisitions went to the Class III depots. Sometimes gasoline distributing points were established at various dispersed localities to service motor vehicles in the air force area.

Class II and IV supplies were obtained in the same manner as Class I and III supplies, except that they were requisitioned only when needed, while the latter were consumed at a more or less uniform rate. Delivery by Quartermaster Truck Companies was the same in all cases. Laundry, shoes, clothing, and articles of equipment needing repair were salvaged and collected at the squadron airdromes, hauled by ration trucks to the service centers, and from there to the depot. Laundry was done by the Laundry Company and returned to the using organizations through the service centers. Shoes and unserviceable articles of clothing and equipment were sent to the Salvage and Repair Company. Replacements were procured from the Class II Depot and returned in the same manner. Articles turned in were classified by the Salvage Officer; those which were repairable were repaired and returned to stock for reissue and the balance was disposed of as salvage.

Service centers were designed to serve two combat groups (eight squadrons). A Quartermaster depot could supply approximately 25,000 troops.[70] One of the functions of a Group Quartermaster in some theaters was to act as Purchasing and Contracting Officer for his area.[71] Unless he used non-appropriated funds, however, he was authorized to buy only such items as would be issued "if they were on stock." If he wanted to buy automobiles, for example, he had to get the Ordnance Officer's certificate that automobiles were unavailable elsewhere and were necessary. Cash for this purpose was obtained from the Finance Officer.

Quartermaster units were first transferred to the Air Service Command for training on 22 June 1942, along with the other arms and services attached to Air Base Groups.[72] There were many types, but the functions of each were distinct. It was felt by the Air Service Command that despite their numbers "no Quartermaster units could be eliminated with impunity." [73] The following Quartermaster units were transferred to, or subsequently activated by the Air Service Command:

Quartermaster Truck Company (T/O 10–517)
Quartermaster Company Service Group (T/O 10–437)
Quartermaster Platoon, Air Depot Group (T/O 10–427)
Quartermaster Company, Depot, Class III (T/O 10–467)
Quartermaster Company, Depot, Subsistence (T/O 10–477)
Quartermaster Car Company (T/O 10–87)
Quartermaster Truck Platoon (T/O 10–518)
Quartermaster Platoon, Transportation, Air Base (T/O tentative)
Quartermaster Company, Medium Maintenance (T/O 10–487) (To Ordnance)
Detachments, 905th Quartermaster Company Service (T/O special)
Quartermaster Rescue Boat Operational Training Unit (T/O special)

The training programs for these Quartermaster units were similar to those of Ordnance and Chemical Warfare, except that no centralized Unit Training Center was ever established for Quartermaster activities either by the Army Service Forces or by the Army Air Forces.[74] The units were stationed at bases scattered over the United States, many of which were not under the Air Service Command. An effort was made, nevertheless, to keep Quartermaster Truck Companies and Ordnance automotive maintenance units together, for the mutual benefit of the two services.[75] There was also a severe shortage of personnel, particularly of officers, in the days before the graduates of the Quartermaster OCS, Camp Lee, Virginia, became available for assignment.[76] Perhaps the most unusual situation in the entire program existed at the school for Quartermaster Rescue Boats conducted by Higgins Boat Industries, which owned 90 percent of the units' operational training equipment (contrary to ordinary War Department policies).[77]

The most important units were the Truck Companies and Platoons, the Service Group Companies, and the Air Depot Group Platoons. The technical training programs for these units were from four to six weeks in duration, and training was conducted under the various air forces and commands. Thereafter the units were assigned to Air Depot or Service Groups for combined training.[78] At the close of the year 1943, approximately 20,000 enlisted men had received specialized instruction in these three main units, and over 140 Truck Companies had been sent overseas.[79] Despite differences in terrain, climate, and methods of combat, the Quartermaster Truck Companies had to haul and protect all types of AAF and ASF supplies that could be found in the several theaters of operations.[80] By the end of the following year, 219 Truck Companies and 50 Air Depot Group Platoons had been sent overseas, along with 40 Boat Crews, 17 Subsistence Companies, 14 Class III Depots, and 1 Quartermaster Car Company.[81]

A number of specialized courses were also conducted by the Quartermaster. These included the following schools: cooks and bakers, mess inspection and management, motor transportation, supply clerks, officers' basic and advanced supply, and marine motors. A unique project, the formation of hydroponics teams for the production of fresh fruits and vegetables by chemical means, was likewise initiated by the Air Quartermaster.[82]

### SIGNAL

Signal Corps [83] organizations in the AAF were charged with the supply, storage, installation where necessary, and salvage of Signal Corps equipment for all units of the air forces. The normal channel of supply was through the Air Service Command, which in turn requisitioned on the Chief Signal Officer, since almost all of the equipment was Signal Corps issue prior to the integration of air communications into the Army Air Forces.[84] Other functions included third and fourth echelon maintenance for all ground and airborne HF, VHF, and FM radio equipment, ground and airborne radar equipment, and wire equipment, in addition to the construction and maintenance of long lines of wire communications between air force installations, when such facilities were not provided by the Services of Supply.[85] Moreover, internal communications systems for outlying bases, service centers, and dispersed air depots were also maintained by Signal Corps organizations. The importance of efficient communications was easy to understand, but in planning theater activities they were often "the last thing thought of and the first thing expected." [86]

In January 1944 three official Unit Training Centers were set up: No. 1 at Robins Field, Warner Robins, Georgia; No. 2 at the ASC Training Center, Fresno, California; and No. 3 at Kelly Field, San Antonio, Texas.[87] The first two specialized in the training of Signal Companies for Service Groups, while the third at Kelly Field continued with the training of Signal Companies, Depot (Avn), as it had been doing for some time. These two units were the most important Signal Corps units associated with the AAF, although several others were trained by the air forces for assignment "one per Air Service Command" or "as required."

A distinguishing characteristic of the Signal Corps was the training of small teams for brief periods of temporary duty in this country and overseas as installation and instructional agents. The training of Signal Companies was insufficient by itself to solve the manifold problems of Signal service; developments in this realm of invention were too rapid. In some cases radio developments required a time lag of only six months from drawing board to operational status.[88] Thus, it was necessary to train military specialists in each new item of equipment and to send them as casuals wherever they were needed, so that they could train the established units in the new techniques of installation and maintenance. When these brief projects were completed, the teams were sent back for instruction in the use of still newer equipment.

As the number of installation and instructional personnel increased, special squadrons or agencies were established to take care of them. Among these were the 1st Radio Squadron, established at Wright Field,

Ohio, on 17 January 1942; [89] the VHF Engineering Agency, Allenhurst, New Jersey; [90] and the 1st Radar and VHF Installation and Maintenance Unit, Aviation, at Warner Robins, Georgia.[91] All three trained and shipped small teams: sometimes only one officer or enlisted man, and other times as many as four or five officers and fifty or sixty enlisted men.

The 1st Installation and Maintenance Unit at Warner Robins was the largest of these installations. It had had a modest start as a refresher school for enlisted radar specialists awaiting assignment to tactical organizations. In the absence of any authorized school, the men had been attached to the local unit of the 859th Signal Service Company (Avn). Much of the program of that unit, and its successors, was basic military training, but eventually, when the equipment became available, extensive courses were set up in airborne radar, airborne radio, beacons, power units, antenna rigging, ground radar, and VHF Fighter Control Systems.[92] The number of trainees varied from 400 in December 1943 to 1,750 in August 1944. An average of 500 men attended the technical courses; others were assigned to Overseas Combat Training Courses or to Non-Commissioned Officers Schools. Graduates were assigned to Signal Companies, Service Group, and Signal Companies, Depot, Aviation, as well as to teams of installation and maintenance personnel.

### Administration of Arms and Services

From the foregoing discussion it should be clear that the air arm in World War II was anything but "separate." The responsibilities of the other arms and services were obviously not inconsiderable. Furthermore, to anticipate a later chapter, these responsibilities were surprisingly permanent. The functions of the various branches were lasting; a few of the training programs described above were continued intact even after the ASWAAF units had become sections within Service Groups (Special), and the integration of arms and services was considered "an accomplished fact." * Not only were the technical sections of the new Service Groups still trained as units, but the component parts of Air Depot Groups, and other separate units, continued to exist in unintegrated form. The indestructibility of the older branches of the army should not be underestimated.

* For further information on these subjects see Chapter 10.

Eventually, however, as the air arm moved toward independence, it did absorb many aspects of the administration of the attached arms and services. Prior to November 1943 responsibility for the training of ASWAAF units rested with special staff officers, who were in the Air Service Command but under the technical supervision of the branches. After the reorganization of the Air Service Command,* special staff responsibilities were transferred to three new functional divisions, and all training activities were taken over by the Personnel and Training Division.[93] At that time the members of the special staff lost their operating functions and retained only their powers to advise, recommend, and coordinate.

For a short time this change caused a loss of efficiency, as Chemical Warfare and Ordnance officers, for example, were obviously more familiar with the problems of their own units than anyone else. There was also a loss of morale, especially during the period between the assumption of control over arms and services and the actual elimination of branch distinctions, for during this period the Air Corps often seemed to ignore the needs of the ASWAAF's. Furthermore the ASWAAF's were cut off from the technical branches of the Army Service Forces; they could now appeal for redress of grievances only to the Army Air Forces. Thereafter, wrote the Chief of the Quartermaster Section,

> . . . When assignments were made by Arms, Area Commanders changed a great many of [the men]. Of course when changes were made the cream of the crop was taken. This section has yet to see one instance when troops have been transferred from the Air Corps Arm to the Quartermaster Arm in order to bolster that Arm up. It has always been in the opposite direction.
>
> What is needed, is education of the various Group and other Commanders to the realization that other Arm and Service personnel is just as important as personnel required for airplane Supply and repairs, and that unless one part of the group is just as efficient and well manned as the other, the group as a whole will be a failure.[94]

Gradually, in spite of these complaints, the advantages of centralized control over supply, maintenance, and training began to assert themselves. The practical improvements engendered by the new system were manifest in the simplification of reporting procedures and the greater ease of manning and equipping all types of units.[95] Since no evidence was ever found of discrimination against ASWAAF's in the

* See Chapter 8.

basic matters of food, housing, recreation, or mail, these improvements in efficiency were ultimately responsible for a corresponding increase of morale.[96]

It was never possible, however, for the Air Service Command to give to the ASWAAF's everything they wanted as a matter of administrative policy. What seemed fair and logical to the several members of the special staff was not necessarily best for the arms and services as a whole. For example, a survey of special staff members in August 1943 revealed that each wanted a proportionate share of all personnel allotted to the Air Service Command, a fair distribution by intelligence levels, as indicated by AGCT scores, and an equal number by SSN in cases where more than one branch used enlisted men with the same classification.[97] Yet it was imperative for the Air Service Command to fill each unit and group in the order of its priority for overseas shipment, and a given branch might have 10 percent of the total authorized personnel, but only 3 percent of the total in committed units. Obviously, therefore, the Air Service Command was forestalled, in the interest of all, from granting the formally expressed wishes of the individual branches. The principle of integrity of command, as it was called, implied a subordination of the branches, and a gradual centralization of control.

★★★ 7 ★★★

# THE PERIOD OF DECENTRALIZED CONTROL

## ASC TRAINING STRUCTURE: 1942–43

FROM 12 DECEMBER 1941 [1] until 1 February 1943 [2] the Air Service Command supervised its many domestic installations through the headquarters of four Air Service Area Commands (ASAC's). These four loosely joined commands coordinated instructions from higher authority and assigned the required projects to control depots and air bases. The power of the four ASAC's within their areas was comprehensive, especially for third echelon work. Fourth echelon work, on the other hand, was always more directly under the central headquarters.

The Commanding General of the Air Service Command until November 1942 was Major General Henry J. F. Miller. He was succeeded by Major General Walter H. Frank.[3] Both of these men were inclined to regard maintenance and supply as the primary functions of their command. At its founding the Air Service Command had been charged with third and fourth echelon maintenance, all types of Air Corps supply, and whatever training was necessary for the accomplishment of these duties.[4] The training program of the Air Service Command was not an exclusively military function; much of it was devoted to the training of civilians for depot and sub-depot positions in this country. Frequently the training of military units was considered secondary to the more immediate problem of operations, or of "production," as it was often called.

Moreover, the majority of service units, especially third echelon units, had to train at bases under air force jurisdiction.[5] Service units had no aircraft of their own; their purpose was to supply and maintain the Fighter, Bomber, Transport, Observation, and other operational groups then being trained by the air forces. To find aircraft and equipment on which to work they had to locate themselves at air force bases. The establishment of "parent" units to coordinate these activities, and to carry on maintenance and supply while trainee units were being transferred from one place to another, was an essential part of the ASC training program. A brief consideration of the scattered and poorly coordinated programs of the four area commands is prerequisite to an understanding of the Air Service Command's efforts to achieve centralization.

The first of the geographical subdivisions was in the northeast quarter of the country.[6] Here was the 1st ASAC, under which, in 1942, Service Groups cooperated with units of the short-lived First Concentration Command, as well as with those of the First Air Force. The operational squadrons of these commands were serviced ordinarily by civilian subdepots; in only a few cases by tactical units. Trainee Service Groups were assembled and given organizational training at Grenier, Westover, Syracause, and Selfridge Army Air Fields for periods of from 60 to 90 days, and then transferred to Fort Dix, New Jersey. At this base the trainee units could understudy the 91st Parent Service Group, which was the original 1st Provisional Service Group.[7]

Normally, trainee units were not sent to Fort Dix until after they had been notified that they were to go overseas. Their stay at this base was to be only a quick check-up of last minute shortages. Requisitions for personnel, submitted at previous stations on AAF Forms 127 and 128, were supposed to have been sufficient to bring the groups nearly to full T/O strength.[8] Indeed, it was anticipated that most of the groups would be overstrength before coming to Fort Dix, for the plan was that they would leave cadres at the stations of organization to assist in the activation of new groups. It was also assumed that upon activation of each new group approximately 20 percent of the necessary organizational equipment would be available. In practice it developed that both expectations were unduly optimistic.

For a short time in the summer of 1942, it was planned that there would be a parent group at Fort Devens, Massachusetts, similar to that at Fort Dix, but because a reduction in the number of Service Groups was contemplated (or rumored), this plan never materialized. Instead, two trainee groups were kept constantly at the Fort Dix Army Air Base. Moreover, cadres were sent not only from the 1st ASAC but from bases all over the country to observe the operations of the 91st Service Group. In the period from 26 June to 23 September 1942, specialized training in the operation of these organizations[9] was given to 1,437 enlisted men and 239 officers, representing 53 Service Groups in almost every state of the union. From the time the functions of Service Groups were separated from those of Air Depot Groups,[10] Fort Dix was a focal point in the training of third echelon units.

Units of the other arms and services were organized and trained by the 1st ASAC at Dyersburg, Tennessee, prior to their assembly at Fort Dix. In this category were Quartermaster Truck Companies, Quartermaster Service Companies, and Ordnance Medium Maintenance Companies (Q). At the same time other attached units were trained at other bases in the area.

The 2nd ASAC in the southwestern section of the United States had an entirely different training structure. A parent group, the 307th Service Group, was located at Will Rogers Field, Oklahoma. Trainee units were organized into groups at that base for a period of from one to three months, during which time they were supposed to receive their organizational equipment.[11] They were then transferred on temporary duty to Woodward, Oklahoma, where they serviced Third Air Force operational training units at five dispersed fields: Hobart, Gage, Clinton, Perry, and Cushing Army Air Bases. After thirty to sixty days of practical experience at these bases, the units were ready for shipment overseas. The main difference between this system and that of the 1st ASAC was that here the activities of the parent group were observed at the beginning of the training period rather than at the end.

The training structure of the 3rd ASAC in the South Atlantic states was the most complicated of all; and the most productive, for this section trained more than half of the service units used in the war.[12] In this area there were three distinguishing features. First, all army air bases engaged in the initial organization of service units specialized in one or two particular types of unit; second, practical experience with combat units was combined with the observation of permanent parent groups during "second phase training," as it was called in this section; and third, parent units at "third phase stations" consisted merely of detachments from the main parent groups.

Prior to each move, cadres were formed and left behind to develop new groups. This characteristic military practice was especially important in the 3rd ASAC at stations of activation and organization, which were located at the following bases:

| | |
|---|---|
| Hq and Hq Squadrons | Dale Mabry |
| Service Squadrons | Hunter, Key, and MacDill |
| Signal Companies | Columbia |
| Ordnance Companies | Dale Mabry, Greenville, Charleston, and Key |
| QM Truck Companies | Dale Mabry, and Key |
| QM Service Companies | Greenville |
| Chemical Companies | Herbert Smart |
| Finance Detachments | Greenville [13] |

Groups were assembled in "second phase training" at Greenville, South Carolina, under the 25th Service Group; at Avon Park, Florida, under the 40th Service Group; at Venice, Florida, under the 27th Service Group. Although the 40th was subsequently moved from Avon Park to Lakeland, Florida, these groups were considered permanent. The turnover of personnel in each was high, however, for individuals and cadres were pulled out with great frequency. At third phase stations there were no permanent organizations, merely detachments from the parent groups. Third phase stations in this area were located at Walterboro, Waycross, Lakeland, Myrtle Beach, and Hattiesburg Army Air Bases.

The 4th ASAC in the Northwest [14] worked in conjunction with the Second and Fourth Air Forces, and in a few cases sent units to Third Air Force bases for combined or third phase training.* First phase training was completed at widely scattered air bases and training centers, where separate squadrons were organized, manned, and equipped prior to transfer to the second phase stations. The latter were established at Pendleton, Oregon, and Muroc, California, where training was conducted under the supervision of parent groups. Ephrata and Walla Walla, Washington, and Wendover, Utah, were also designated as second phase stations, although they were not under the parent group system. Only rarely were Service Groups assigned to air force stations to operate without the assistance of parent groups, or without prior training as assembled organizations.

Cadres for new Service Groups were filled and trained at Pendleton, Oregon, under the 330th Parent Service Group, and at Muroc, California, under the 31st Parent Service Group. When complete cadres had been formed at these bases, they were transferred to other stations to as-

* The three phase training plan of the 3rd ASAC eventually became general. The terminology, however, was picked up before the practice was fully introduced.

sist in organizing and training new tactical units. After thirty to sixty days at second phase training stations, Service Groups were moved to third phase stations at Great Falls, Montana, Rapid City, South Dakota, and Sioux City, Iowa, for combined training.

Early in 1943 a centralized first phase station was established at Santa Maria, California, where most of the units assigned to Service Groups were assembled, equipped, and manned for assignment to second phase stations. This was done to eliminate irregularities previously encountered when incoming units reported at second phase stations short of personnel and equipment, and lacking the military and technical background to carry on group training.

Fourth echelon service units in this period were likewise trained by the four ASAC's, but ordinarily they were not moved about with such rapidity. The usual procedure was to activate Air Depot Group units at control depots and then to send them for second phase training to air depot training stations closer to the flying training activities of the air forces. At each air depot training station there was a full set of organizational equipment, which had belonged originally to the first group ordered to that base and had been left behind as a permanent contribution to the local facilities. Departing groups took with them the newly packed equipment of the groups assigned to fill their place.[15]

In December 1942, the Acting Chief of the Personnel and Training Division, ASC, recommended that the Air Service Command be authorized to carry at least one Air Depot Group at cadre overstrength at each of the following stations: Albuquerque, New Mexico, Stinson Field, Texas, New Orleans, Louisiana, Wellston, Georgia, Charleston, South Carolina.* This recommendation was made because there were no parent units for Air Depot Groups, and it was believed desirable to provide cadres for the activation of new groups. Further justification was the expectation that the Air Service Command would be called upon to provide extra Repair and Supply Squadrons, not included in the AAF Program for 1943.[16]

The plan, however, was disapproved on the ground that the extra Repair and Supply Squadrons were to be activated only at the rate of four per month until a total of ten had been obtained, and also because there was already a surplus of Air Depot Groups.[17] In other words, the

* Other important air depot training stations were located at Springfield, Ill., Oklahoma City, Okla., and Hensley Field, Texas.

Air Service Command was ahead of the air forces, so far as fourth echelon units were concerned, and should now concentrate on quality rather than quantity. Nevertheless, the command was instructed that "when and if the number of Air Depot Groups considered surplus becomes five or under, the above request [to hold groups at cadre overstrength] would be pertinent and should be re-submitted."[18] This provision became important in 1943 as a result of increased requirements for the Air Transport Command and the militarization of sub-depots.[19]

The management of all these units was an adaptation of the long-established War Department policy for the activation and development of Infantry divisions. Units were constituted on paper by War Department authority, and activated by appropriate headquarters upon receipt of that authority. The process of activation involved the assignment of personnel and equipment in accordance with the designated War Department tables. When units were no longer needed, they were inactivated (the term deactivated was also used) or disbanded. In the former case they were reduced to zero strength, or almost to zero strength, but were still available for movement, less personnel and equipment, either in the United States or overseas. In the latter case designations, personnel, and equipment were absorbed by other units or discontinued altogether. Once a unit was disbanded, it could be used again only if it were reconstituted by War Department authority. In this way the War Department retained a certain amount of control over the activities of many heterogeneous headquarters.

### Surplus of Personnel

At the beginning of 1942 there was a shortage of military personnel in all AAF commands, because so many units were activated immediately after Pearl Harbor. Later in the year, personnel were inducted faster than either the training organizations or the tactical units could take care of them. During this short period, there was an actual surplus of personnel in the Air Service Command and in some of its attached services—a unique situation which was not repeated until much later, when the war was almost over. Unfortunately, in the early period many of the personnel were classified as "limited service" and were unqualified for overseas duty. In spite of the surplus within the command, therefore, there was never a surplus for tactical units.[20]

Training was definitely affected by both shortages and overages. The system of allotment was as follows: once every week, on Friday, the Army Air Forces received a teletype from the Technical Training Command which contained the number of available unassigned Air Corps personnel, by specialty, who would complete their courses at basic training centers in the second calendar week following the submission of the report; on Wednesday the Army Air Forces received a similar report from the special service schools of the Services of Supply (Army Service Forces).[21] Following receipt of these reports, Headquarters, AAF, informed its subordinate commands of the trained personnel available from all sources, and in turn received requisitions from these commands in accordance with their requirements. Final assignment was made by the Army Air Forces at the direction of the War Department. Personnel eliminated from these courses for any reason were transferred to the nearest of the following AAF basic replacement training centers:

Keesler Field, Mississippi (White)
Miami Beach, Florida (White)
St. Petersburg, Florida (White)
Atlantic City, New Jersey (White)
Sheppard Field, Texas (White)
Kearns Field, Salt Lake City, Utah (White)
Fresno, California (White)
Jefferson Barracks, Missouri (Colored)[22]

The flow of trained personnel to the Air Service Command was directed insofar as possible to service units committed for overseas duty.[23] The number of specialists required by these units, however, was large and the number of Technical Training Command and Services of Supply schools was not. Consequently, it was necessary for the Air Service Command to utilize every available resource and facility to supplement this flow and to obtain more trained personnel. This was done by conducting recruiting campaigns to procure experienced mechanics and technicians, and by making determined efforts to establish additional special schools.

The success of these efforts was attested by the fact that in the latter part of 1942 and the early part of 1943 the Air Service Command had, as has been pointed out, a large surplus. On 6 February 1943 there was a surplus of 21,019 men, all Air Corps personnel, based on current strength as authorized by The Adjutant General.[24] Although future needs would quickly absorb these men, the situation was temporarily embarrassing to the Air Service Command, for there were no centralized "replacement wings" to handle the men until they were assigned to tactical units, and the command was criticized for idle surpluses. "Informal information," wrote Major General Davenport Johnson, "indicates that filler personnel received by your command are being held in large pools, and not being filtered in service group training centers without delay." [25] This criticism was made in spite of the fact that many of the men were unqualified for overseas service.

One of the complications, according to Brigadier General Elmer E. Adler, was that "this command cannot assign surplus personnel to the majority of its organized units, as such units are housed at bases of other commands; such housing being under the control of the command concerned. These commands object to units being over T/O strength, and are requesting that such surplus be transferred." [26] Consequently, for several months, Station Complements (later Base Headquarters and Air Base Squadrons) and similar units located at depots and other permanent ASC installations were swamped with surpluses, with a disastrous lowering of the morale of the men, for as General Adler pointed out, "surplus assignment is a contributing factor to malassignment, in that excess personnel are generally employed in duties not in harmony with their training." [27] Furthermore, in overstrength units there could be no opportunity for promotion.

Centralized replacement wings (or replacement depots), to correct these evils were not set up until the fall of 1943. Meanwhile, a number of units were authorized by The Adjutant General to retain personnel in excess of established T/O's: Station Complements for Air Corps personnel, and a few tactical units for other arms and services. Officially these units were authorized only 33⅓ percent overstrength; [28] actually in some cases they were 1,000 percent overstrength, notably at Warner Robins, Georgia; San Antonio, Texas; Fairfield, Ohio; and San Bernardino, California. According to regulations, assignment and reassignment to and from these units was to be made only by the War Department, through The Adjutant General, and a full report was to be rendered monthly by The Adjutant General to G-3 (Training and Operations) of the War Department with complete details as to the status of all replacement units.[29] In practice this procedure applied only to the initial allotment or assignment to the various replacement units. Thereafter the Air Service Command had informal authorization to distribute personnel as it saw fit.[30]

### The Parent Group System

When Air Base Groups were transferred to the Air Service Command in June 1942, and redesignated Service Groups, the decision was made to take the training of third echelon units away from the numbered air forces, and to delegate both third and fourth echelons to the Air Service Command. But no matter which command was in charge, third echelon service units had to be trained with the operational units of the air forces. From this situation arose the necessity for parent groups, a system of instruction (and also of supply and maintenance) which was a compromise between the Air Service Command and the air forces, destined to last until the final weeks of the war.

Dual command aggravated the problem of "production vs. training," for air force commanders were interested in their own training programs —in flying training—and demanded the maximum number of serviceable ships at all times.[31] They were anxious to have the most efficient service they could get, whether it was performed by depots, sub-depots, civilian overhaul establishments, or by tactical groups, and usually they avoided trainee units as the least efficient agency of all. "I've got a job to do," they would say, "and I don't like to have a lot of green men monkeying with my equipment, because it will slow me down." [32] Furthermore, mechanical equipment was at a premium, and the air forces did not want it used for unproductive training.

The system of instruction thus forced on the Air Service Command meant in the case of Service Groups that a parent group (over one thousand men) would be stationed at an air force base to service an operational training unit then in flying training or on maneuvers, while a trainee unit (another one thousand men) would be stationed at the same base from one to three months to "observe" the activities of the parent group. From a pedagogical standpoint, this system had certain obvious drawbacks, but it was the logical and inevitable result of divided responsibility.

Many officers (and civilians) in the Air Service Command were anxious to alter or eliminate the whole parent group system. A more economical proposal, these people felt, would be the establishment of small training teams, with experts, both officers and enlisted men, qualified as instructors in their own particular specialties. One of these teams would be assigned to supervise and direct the training of each new Service Group. By this means more effective training would be achieved, and the waste and duplication of the parent group system would be cut "to the vanishing point." [33]

Furthermore, the original plans for the disposition of Service Groups had undergone two changes. First, combat groups in the United States had been operating primarily from group airdromes, rather than squadron airdromes, so that Service Groups could not deal with dispersed units, as contemplated in their basic regulation.[34] Second, these group airdromes had been situated at fixed installations, with Base Headquarters and Air Base Squadrons, Guard Squadrons, sub-depots, and all the necessary Station Complement personnel. Thus Service Groups, even in some cases parent Service Groups, had to occupy small portions of the airdromes and "scramble for what work and training might be left over." [35]

For these reasons General Frank recommended in February 1943 that all final phase training stations be removed from the jurisdiction of the air forces and assigned to the Air Service Command.[36] More specifically, he requested that all Base Headquarters and Air Base Squadron and Station Complement personnel, with the exception of the necessary Guard Squadrons, be removed from these bases, and that Service Groups in final phase training be permitted to operate the sub-depots and furnish all third echelon maintenance, supply, and administration. Objections to the Air Service Command's controlling these bases, General Frank admitted, "would no doubt be many—emphatic, and in some cases logical." But it was his opinion that this plan presented the best possible procedure, not only for the improvement of the training of Service Groups, but also for the education of combat commanders as to the functions and duties of their supporting organizations. Once command control had been assumed over all final phase training stations it would not be difficult to effect sweeping reforms in the parent group system.

Unfortunately this request was refused. The plan was referred to the Commanding Generals of the Second and Fourth Air Forces, and Headquarters, AAF, decided that two compromises were necessary:

a. The command of the Air Base involved must remain with the Air Force concerned.

**b. In order to maintain continuity and quality of service, a parent group must be established at the base concerned.[37]**

Plans to revamp parent groups died with this defeat; the system was continued for more than two years.[38] Not until the air forces took over all sub-depots of the Air Service Command, on 1 January 1944,[39] were air force bases in a position to handle their own third echelon maintenance and supply without dependence on parent groups, and even then no change was made. Finally, in the early months of 1945, parent groups were discontinued and the long hoped-for training teams were instituted.[40]

An illustration of the confused conditions inherent in the parent group system can be found in the surfeit of technical regulations issued by the various headquarters to their subordinate units. Trainee organizations, at the bottom of several chains of command, were expected to keep track of every regulation published by every headquarters all the way up to the War Department. To simplify matters, each headquarters rewrote the directives of the headquarters immediately above, but this process only increased the total output. There may have been other reasons for these conditions in some places, but the result, so far as trainee organizations were concerned, was an impossible burden of directives. For example, squadrons trained over a three-year period by the 40th parent Service Group at Avon Park, Florida, were instructed to maintain at all times complete and up-to-date files on the following (in addition to a large roomful of Technical Orders):

| | |
|---|---|
| Army Regulations | Mobilization Regulations |
| WD Circulars | Technical Manuals |
| WD Memorandums | Field Manuals |
| AAF Regulations | WD General Orders |
| AAF Memorandums | WD Bulletins |
| Third Air Force Memorandums | Base Memorandums |
| ASC Regulations | 40th Service Group Memorandums |
| Warner Robins ASC Regulations | Special Orders |
| ASC Memorandums | General Correspondence |
| Training Circulars | Squadron Orders, etc.[41] |

### Air Depot Groups

A quite different approach was needed for the training of fourth echelon units. The nature of fourth echelon work made it impossible for the component parts of Air Depot Groups to be trained at outlying bases; they had to be trained at special air depot training stations and at air depots. The management of these units was less a matter of coordination with the air forces than of obtaining the required equipment and utilizing the facilities of established civilian installations. For this reason it was not so necessary to institute parent Air Depot Groups, although one such group was set up in December 1943.[42] The object was to arrange for the performance of fourth echelon work without interfering with the commitments of the depots: the familiar conflict of "production vs. training" but without the complication of dual command. Mostly it was a matter of internal administration, of setting up schools and special courses, and providing for the classification, distribution and assignment of personnel. The manner in which this was done can be illustrated by a few specific examples.

### AT WARNER ROBINS AND MOBILE

The 4th and 5th Air Depot Groups had been stationed at Herbert Smart Airport in the Warner Robins area during the Carolina maneuvers of 1941. In February 1942 three new Air Depot Groups were established at the same location. Fillers for these groups, the 36th, 37th, and 38th, were furnished after their arrival in Georgia. The 36th arrived first, and 235 enlisted men were transferred thereto from the 5th, which was still at the Herbert Smart Airport. Fillers for the other groups were dispatched to the same field from Jefferson Barracks, Missouri, and Keesler Field, Mississippi.

Incoming personnel were assigned to one or another of the three groups and then transferred to civilian schools for technical training. The intention was to have about 500 men absent at one time,[43] but only 280 in all were actually sent.[44] The schools were located in all parts of the country: the Embry Riddle School of Aviation, Miami, Florida; the Frank J. Ambrose Aviation Institute of Technology, Long Island, New York City; the Lansing Section of the Ordnance School, Wells Hall, Michigan Stage College; the Coine Electrical School in Chicago, Illinois; the Curtiss Wright Propeller School at Caldwell, New Jersey; the General Electric Supercharger School in West Lynn, Massachusetts; and the Delta Air Corporation in Atlanta, Georgia. Courses were taken at these schools in engine overhaul, sheet metal, welding, electrical work, and various accessories. In addition to this personnel, the three groups received a total of 665 men who had been trained at such service schools as the Air Corps Technical School at Chanute Field, Illinois.[45]

Living conditions at the training base left much to be desired. Herbert Smart Airport was essentially a tent colony, and the monthly housing and strength report for March 1942 indicated 260 tents, housing 678 enlisted men and 24 officers. Because of the lack of messing facilities, the men of the 36th, 37th and 38th were attached temporarily to the 5th Air Depot Group for rations. When the 5th left, the men of the 37th and 38th were attached to the 36th for rations. Troops were hospitalized when necessary at the Station Hospital of Camp Wheeler, a neighboring Infantry Replacement Training Center, since the buildings at the Wellston (Warner Robins) Air Depot were still in the process of construction.[46]

At first it was not considered necessary to have an administrative headquarters, but personnel continued to increase, and on 11 April 1942 the 4th Station Complement was activated (along with ten others in the country),[47] and the morning report for that date showed "1 EM asgd and jd." On the last day of April, Lieutenant Colonel Russell Scott, subsequently Deputy Commander of WRASC, was designated Commanding Officer of Herbert Smart Airport and also of the 4th Station Complement, which by that time had increased to a strength of fifty enlisted men assigned, and twenty-five attached. By midsummer, facilities at Robins Field had been completed, and on 16 August 1942, the 4th Station Complement was moved with all of its personnel, files, and equipment, into the Cantonment Area of the new depot. The day before it was transferred, the 20th Station Complement was activated at Herbert Smart Airport to replace it.[48]

The significance of these small beginnings was that Station Complements were to serve as housekeeping units, training structure units, and classification centers for Air Depot Groups during their initial periods of organization.[49] The 20th became the housekeeper for the 38th and 46th at Herbert Smart Airport, and the 4th acted in the same capacity for the two groups after they had been transferred to Robins Field. Personnel of the latter unit also helped with the 33rd, 40th, 55th, 58th, and 64th Air Depot Groups, which were stationed at Robins Field during 1943.

The growth of the new training center at Robins Field was so rapid that there was little opportunity for planning. A request was made for an increase in the authorized strength of the 4th Station Complement, "in order to facilitate the handling of up to 300 casuals." It was informally granted, and the 4th soon grew from 42 enlisted men to nearly 400. From this situation sprang the characteristic "pool for unassigned personnel" in which men were carried as casuals, initially for purposes of classification, and later because they were surplus and there was nothing for them to do. Only those with specific jobs at the depot could feel that their time was profitably employed. The plight of the local commander was described by Colonel Russell Scott, Executive Officer in Charge of Troops, as follows: "This Headquarters . . . receives a large number of men who are assigned to units that have not been activated at this station, to units not at this station, or to units that have left this station. This has caused a great deal of confusion." [50]

The facilities at Warner Robins, nevertheless, were excellent and in January 1943 Colonel Leo H. Dawson, a roving inspector from Headquarters, Air Service Command, reported that an "ideal training spot" had been created at the Warner Robins Air Depot. In addition to the newly organized school for officers, and the centralized training shops, there were three dispersed Air Depot Group areas, at which a maximum of training could be accomplished with a minimum of interference with the productive activities of the depot. This was an unusual advantage. In Colonel Dawson's opinion, "it was the best set-up in the continental limits . . . for complete Air Depot Group training," and he recommended that it be kept fully employed at all times.[51]

Colonel Dawson was well qualified to express an opinion, for at the time of his recommendation he had visited training centers at Albuquerque, Oklahoma City, San Antonio, San Bernardino, Sacramento, Ogden, New Orleans, Mobile, and Charleston. Altogether, he had conducted formal inspections of fifteen groups, "some of which were reinspections of groups that were not considered satisfactory on the initial inspection." [52] During this time, he had served as ambassador plenipotentiary for the Air Service Command with power to make or break the Commanding Officers of the bases or groups inspected. On rare occasions he had relieved inefficient officers on the spot; ordinarily he had made recommendations to Headquarters for appropriate action, and always his recommendations were supported.

At regular intervals Colonel Dawson reported his findings to Brigadier General Clements McMullen, Chief of the Maintenance Division, and later to Brigadier General E. E. Adler, Chief of the Personnel and

Training Division, Air Service Command, describing conditions base by base and unit by unit in carefully written longhand letters. His conclusions, in view of his experience, were noteworthy and encouraging, for he reported a marked improvement over a three-month period. "When I first started inspecting Air Depot Groups," he wrote, "all of the Control Depot Commanders were quite adverse [*sic*] to utilizing the personnel of the Air Depot Groups in their shops. This resistance was occasioned by the feeling of the respective Depot Commanders that the employment of the Air Depot Groups personnel would retard the production of the Depot. Their resistance has been broken down and all of the Depots are now employing the personnel of the Air Depot Groups under their command to the maximum degree." [53]

The reported change in attitude had helped to improve the training of Air Depot Groups, and also to speed up the production of the depots. In this connection, the situation at Mobile, Alabama, was cited by Colonel Dawson.[54] At that depot, personnel of the 44th and 45th Air Depot Groups had taken over the activities of the third or "graveyard" shift, and had actually produced more than the civilians on the other, more favorable shifts. On one occasion, twelve engines had been torn down at night by military personnel, as compared with nine by civilians on the following shift; five superchargers had been reconditioned, as compared with three by civilians. Throughout the shop the same general average had been observed at other times. In the supply depots, he noted, military personnel had been more careful than the civilians in crating and marking supplies for overseas shipment. "This result had been occasioned," Colonel Dawson believed, "by constant harping on the subject by the undersigned and by repeated inspections." Unfortunately, the problem of production could not always be solved by assigning military personnel to the depot's third shift.[55]

### AT SAN ANTONIO

Here the training of Air Depot Groups had taken much the same form as at Warner Robins and Mobile.[56] At San Antonio, as at Warner Robins, two of the original units had been organized in 1941 with experienced, Regular Army personnel. There was also a similar connection between the earlier groups and the newer ones, for the 3rd before its departure had provided cadres of 35 men each for both the 8th and the 9th. The latter two groups, augmented in February by 500 selectees and at irregular intervals thereafter by graduates of the Air Corps Technical Schools, furnished cadres of as many as 77 men for the 12th, 13th, 14th, 15th, 16th, and 17th.

About March 1942, a Training and Liaison Office was established in the headquarters of the San Antonio Air Depot to supervise the administration and training of all Air Depot Groups in the area.[57] When, on 20 May 1942, the 6th Station Complement was activated, it was placed under the Training and Liaison Office. But the authorized overhead of the new unit was only 6 officers and 42 enlisted men—for Medical, Finance, Ordnance, Signal, Quartermaster, and Military Police detachments. Furthermore, all military personnel entering the area were assigned to the 6th for classification and subsequent transfer, and control of these transfers was exercised not by the San Antonio Air Depot, but by the 2nd ASAC with headquarters at Fort Worth.[58]

At times the 6th Station Complement was greatly overstrength. On 26 September Colonel Paul C. Wilkins, in command of the depot, requested an increase in the T/O of this unit on the ground that six Air Depot Groups were then in training, and the total military strength of these groups was as high as 5,000 men. Moreover, the Duncan Field training zone lay two miles from the depot headquarters, and required an entirely separate administration. Training was continuous, twenty-four hours a day, in three eight-hour shifts, and it was not considered feasible to draw on trainee personnel to accomplish base duties. He proposed a minimum of 18 officers and 130 men.[59]

Colonel Wilkins' request was refused, the Office of The Adjutant General stating that a Base Headquarters and Air Base Squadron was contemplated for Duncan Field, and that additional personnel would be authorized when the new unit was established.[60] Meanwhile, enlisted men continued to arrive at an unprecedented rate. By the end of October 1942 the Commanding Officer of the San Antonio Air Depot reported, in a request for WEMA (recreation) funds, addressed to the 8th Service Command at Fort Sam Houston, Texas, that the potential average of unauthorized casuals attached to the 6th Station Complement was 2,500. This condition lasted until 9 January 1943, when the 481st Base Headquarters and Air Base Squadron was activated and officially designated as a Replacement Center for Air Corps troops. The 6th Station Complement was disbanded at that time, and all personnel were absorbed by the new unit.[61]

Also activated at Duncan Field were the 7th Station Complement, which was later transferred to Albuquerque, New Mexico, and the 16th Station Complement, which was sent to Stinson Field, not far from the city of San Antonio.[62] Both of these units were likewise supplanted by Base Headquarters and Air Base Squadrons; by the end of 1942, Stinson Field was one of the busiest air depot training stations in the country. Another such installation was located at Hensley Field, Dallas, in August 1942.[63] Until June 1943, however, the training there was limited to one type of unit, the Quartermaster Truck Company (Avn), which consisted in this case of Negro troops.

On 14 May 1942, the 9th Air Depot Group was transferred from Duncan Field to an east coast Port of Embarkation. The 8th departed for the same destination on 16 August, and other groups followed in rapid succession. Instead of packing and crating their equipment, including assigned vehicles, and taking it all with them, as the 3rd and the 5th had done, these groups were to receive their impedimenta at an Intransit Depot just before they embarked.[64]

Meanwhile, significant changes were taking place in the management of on-the-job training at the San Antonio Air Depot. Originally, military personnel had been subject to the same rules and regulations as civilians, and had had a civilian instructor for each ten men. Under this regime, as it turned out, a good deal of jealousy had developed, and the soldiers had tended to idle and to claim special privileges because of the disparity between civilian wages and military pay and allotments. Civilian instructors lacked authority over their students, and "absenteeism" became a problem. Attempts were made to ameliorate these conditions by relieving the military of all duties outside the depot, and by placing officers and non-coms in charge at all times, but the trouble was not altogether eradicated. In addition, it was found that the gap between depot and group practices was so wide that it was practically impossible to present the relevant aspects of maintenance and supply in a way that would be profitable to members of tactical units.

To eliminate incidents arising from the mixing of civilians with the military, the entire third shift of the depot was turned over to the groups. But here as at Mobile, this policy created difficulties of its own, for the work of the third shift was limited both in extent and in variety. The men were unable to obtain a sufficient number of crashed aircraft, and even when they could get the planes, they were unable to perform the entire cycle of work. The next shift took up where they left off; they finished only the smallest of jobs. The depot was unable to simulate conditions in the field.[65] It was only with the establishment of the previously mentioned air depot training stations that this ideal began to be realized. When the formal three-phase training system was inaugurated in all areas, still later, the program began to function with gratifying success.

★★★ 8 ★★★

# DEVELOPMENT OF A UNIFIED PROGRAM

## ASC TRAINING STRUCTURE: 1943–44

When Major General Walter H. Frank was assigned to the Air Service Command in November 1942,[1] a number of sweeping changes immediately took place. On 8 December 1942 Headquarters, ASC, was transferred from Washington, D.C. to Dayton, Ohio, and effective 15 December it was completely reorganized.[2] The chief purpose of General Frank's innovations was to bring to the headquarters of the Air Service Command the principle of industrial, or functional, organization rather than the obsolete straight-line military organization represented by the familiar G-1, G-2, G-3, and G-4. The Personnel and Training Division was set up at this time (along with Maintenance and Supply) as one of the three operating divisions in the headquarters, and shortly thereafter Brigadier General Elmer E. Adler was appointed its chief.

On 1 February 1943, the four Air Service Area Commands were disbanded and eleven Air Depot Control Areas were established instead.[3] The new headquarters were located at, and assumed command over, the eleven control depots in the United States, but they served also as area commands with jurisdiction over all ASC installations and units in territories ranging from two to a dozen states. Under this setup, it was more than ever necessary for Headquarters, ASC, to develop a unified training program that would standardize procedures throughout the country, and assure coordination of the several depots.

The impetus for the new training structure, however, came originally from Headquarters, AAF, which was already concerned with the blocking out of shipments for the European invasion. On 31 January 1943 Brigadier General Lyman P. Whitten, Director of Base Services, Headquarters, AAF, wrote to the Air Service Command with reference to the AAF Training Program, as discussed at a recent conference in Washington, and requested a complete and detailed plan for air service units. The plan was to include a list of the parent groups, the number of proposed stations of activation, the names of second and third phase training stations, and the designations of filler units to complete the groups then overseas.[4]

Brigadier General E. E. Adler, the new Chief of the Personnel and Training Division, ASC, held several conferences with members of his staff to discover what obstacles had been met in the past, and in March 1943 directed the Chief of the Training and Operations Section, Colonel Dwight B. Schannep, to submit to him a program with the agreed-upon solutions. In making this request, General Adler outlined his own understanding of the most important points to be considered.[5] The training of Service Groups, he said, must be planned so that these groups "will progress normally through the first, second, and third phases of training, gaining their equipment as they progress, and finally operating under simulated field conditions." [6] Air Depot Groups "could best be trained in the depots." If opposition arose to the presence of soldiers in the depots, some compromise would have to be worked out, such as the establishment of training areas adjacent to the depots, as at Warner Robins.

Colonel Schannep started his study with Air Depot Groups. Although fourth echelon work was more technical than third, the training of groups to accomplish it was simpler to organize, for they were trained exclusively at ASC installations.[7] The training period was to be divided into three phases: "basic" or first, "technical" or second, and "organizational-field" or third phase training. During the first or basic phase all personnel were to be classified, assigned, and organized. Such administrative details as immunization to disease, the accomplishment of service records, and the issuing of individual equipment were to be completed during this phase, and the following subjects were to be taught: basic drill, range fire, administrative training, physical conditioning, discipline, and indoctrination on the articles of war and sex morality. Second phase training, the training of individuals in technical specialties, was to be accomplished at air depots. Unit training was to be given there by "experienced instructors" and with "efficient machines," when available. Third phase or organizational-field training was to be given under combat conditions with emphasis on housing, messing, camouflage, and the physical toughening of all personnel. An ample supply of "modern repairable wrecked combat-type airplanes, engines, motor vehicles, instruments, radios, etc." would be provided, and personnel would have the "same tools which units will have in performing jobs overseas." [8] Stinson Field, Texas, New Orleans, Louisiana, Albuquerque, New Mexico, and the Cantonment Area of Robins Field, Georgia, were particularly recommended as third phase training stations for Air Depot Groups, with additional groups at the AAF School of Applied Tactics, Orlando, Florida, and the Desert Training Center in southeastern California.

Colonel Schannep felt that technical school graduates needed the experience of third phase training to gain proficiency in their specialties. He also felt that regardless of previous instruction by the Technical Training Command, additional practice in the required crafts was needed at the depots. "In the fingered dexterity of actual manipulation," he said, "the technical school graduate has never had sufficient time to gain efficiency." [9] Also, he declared, Air Depot Groups should start second phase training at full T/O strength for both officers and enlisted men, and credits should be set up for supplying the groups with both AAF and ASF equipment at the beginning of the training period; it should not be necessary to wait for warning orders to obtain priority. As an indication of the relative importance of the various units of an Air Depot Group, he stated that requests from the theaters demonstrated a disproportionate demand for Depot Repair Squadrons, and a slightly smaller demand for Depot Supply Squadrons. There had been no calls whatever for arms and services units, or for Headquarters Squadrons. Accordingly, he recommended that ten additional Repair Squadrons and five additional Supply Squadrons, not connected with groups, be activated immediately. This last suggestion was acted upon early in April.[10]

The geographical distribution of Air Depot Groups came in for particular criticism from Colonel Schannep. "Training," he said, "should be concentrated in as few places as possible so as to promote standardization and efficiency. The idea of scattering Air Depot Groups throughout the U.S. is not efficient." [11] Middletown and Spokane Air Depots should not be used; Rome and Sacramento should be employed only if absolutely necessary; and concentration of Air Depot Groups at existing facilities should be as follows:

| | |
|---|---|
| Oklahoma City Air Depot | 2 groups |
| Ogden Air Depot | 3 groups |
| San Antonio Air Depot | 3 groups |
| San Bernardino Air Depot | 1 to 2 groups |
| Fairfield Air Depot | 3 groups |
| Warner Robins Air Depot | 3 groups |
| Mobile Air Depot | 2 groups |

To plan the training of Service Groups, a conference was called at Patterson Field, Ohio, on 9 and 10 April 1943, which included representatives of the four continental air forces and the eleven ASC Control Area Commands.[12] The inclusion of the representatives of the air forces was necessitated by the previously mentioned failure of the Air Service Command in March 1943 to obtain control over all final phase Service Group training stations, and by the decision of Headquarters, AAF, that parent Service Groups would have to be maintained by the Air Service Command at bases under air force jurisdiction.[13] As a result of the April conference, and of Colonel Schannep's previous recommendations, a revised training program was drawn up, and on 20 April 1943 it was presented by General Adler to General H. H. Arnold, Chief of the Army Air Forces, and Major General O. P. Echols, Assistant Chief, Air Staff, Materiel, Maintenance, and Distribution, at Headquarters, AAF.[14]

General Adler's presentation contained a complete outline of the various phases of training for all ASC organizations, both Air Corps and other arms and services. It contemplated a seven months' training period for Service Groups and a six months' training period for Air Depot Groups, and it was based on the assumption that eight new Service Groups and four new Air Depot Groups would have to be readied each month. Three phases of training were to be provided for both types of groups; the component units were to move from station to station with only their housekeeping, or Quartermaster, equipment. Complete sets of other organizational equipment would be permanently spotted at the designated training stations.[15]

Parent Service Groups, General Adler said, were to be located at second phase Service Group training stations, and were to be held constantly at a three-cadre overstrength. Each parent group was to train one extra cadre per month, and was to receive a constant stream of fillers for that purpose. Thus, parent groups were responsible not only for the organization of cadres, but also for the supervision of fillers, both commissioned and enlisted, for several months prior to the activation of the units for which they were intended. It was expected that cadre officers would spend a great part of the preactivation period in the Command and Staff School at Warner Robins, Georgia, and the AAF School of Applied Tactics, Orlando, Florida.[16]

At third phase Service Group training stations Parent Service Groups (Reduced), the forerunners of Service Squadrons (Training), were to be stationed to insure continuity of service when one trainee group moved out and another moved in. These reduced groups were also responsible for the training of new organizations, but they were not required to furnish cadres. Sub-depots at second and third phase stations and at all satellite airdromes thereof were to be inactivated. Base special staff functions, including Quartermaster, Ordnance, Signal, Medical, and Finance, were to be taken over by Service Groups, but other base functions, such as accountability for ASWAAF supplies, were to remain with the base commands.

Air Depot Groups, General Adler stated, had no need for parent organizations. Instead, each trainee Air Depot Group was to be built to a one-cadre overstrength during its first two phases, and at the end of this period the cadre was to be withdrawn for the activation of a new group. Officers and enlisted men were to be trained and ordered to schools in the same manner as prescibed for Service Groups. With one exception the stations selected for the final phase training of Air Depot Groups were so situated that they could be utilized to furnish fourth echelon supply and maintenance for two Service Groups, as required by AAF Regulation No. 65–1; that exception was Stinson Field, San Antonio, Texas, where tactical groups were performing a portion of the work of the San Antonio Air Depot. In that one spot the arrangement was so successful that no change was contemplated.

At the conclusion of his presentation, General Adler pointed out that action on the part of higher authority was definitely indicated. In particular, he recommended that the appropriate staff offices of Headquarters, AAF, be directed to perform three things: (1) provide for the assignment of filler personnel within thirty days of the activation of each new unit, so that the training of all unit personnel would begin simultaneously; (2) provide for the movement of completely trained units out of the final phase training stations, so that the orderly flow of the program would not be stopped by the irregularity of shipping schedules or by other logistical arrangements which had nothing to do with the ultimate demand for service units; and (3) authorize the activation of Parent Service Groups (Reduced) to meet the needs of the final phase stations in the Service Group training program.[17]

General Arnold approved the ASC plan, as presented, with one or two qualifications: for example, he expressed concern over the time element, and directed that first phase training be reduced from three to two months for all units. He also noted that the plan "requires too much moving." The difficulty here was that whenever any moves were eliminated, the number of service organizations at a given base increased to the point where there was insufficient work to go around.[18] This dilemma was pointed out to him, and accordingly the problem was left in abeyance for the time being, but on 30 April 1943 the official AAF decision was handed down by Major General O. P. Echols: "First and second phase training of Service Groups should be conducted at one station as is done in the case of Depot Groups. This can be accomplished by providing 16 first and second phase stations in lieu of 8 of each. This change would provide the additional benefit that a station would have to provide a cadre for a new group every other month in lieu of every month."[19]

It was also suggested that "insofar as possible Service Groups should be completely trained within the same Depot Control Area during all three phases in order that responsibility can be definitely fixed for its training."[20] Another alteration directed by higher headquarters, the execution of which turned out to be impracticable, was to have parent groups take over complete responsibility for all base activities, including accountability for supplies. It was thought that this change would result in economy of manpower and elimination of the conflicts of divided responsibility.

In view of the urgent need for trained service units, however, it was understood that these reservations were not to retard the program, but rather that undesirable features would be eliminated as the plan progressed. Even before written clarification of the questionable points was received, General Adler made a "tour of coordination" to the areas (from 28 April to 9 May) to give "personal impetus" to the new arrangement. Meanwhile, General Echols wrote to certain staff members at Headquarters, AAF, to urge prompt action. "It is desired to point out," he wrote to the Assistant Chief of Air Staff, Personnel, "that the present plan of furnishing units 25% of their personnel at the time of activation and 25% additional each month thereafter until filled up has caused considerable trouble in filling organizations, and has retarded training." [21] A separate communication to the Assistant Chief of Air Staff, Operations, Commitments, and Requirements, was forwarded on the same date to obtain authorization for Parent Service Groups (Reduced).[22]

On the basis of verbal agreements, and General Echol's subsequent confirmation, the Air Service Command issued a new directive to all area commanders on 1 June 1943, entitled "Revised Training Program for Service and Air Depot Groups." [23] This directive was the first to present a unified, national training structure for all service units. It was based on the hypothesis that eight new Service Groups and four Air Depot Groups would be required each month. There were to be eight separate systems, or flows, of units for each of which there would be an "expediter" at Headquarters, ASC. The flows were arranged as indicated in Table 9.[24]

TABLE 9

THREE PHASE TRAINING PLAN FOR SERVICE GROUPS, 1 JUNE 1943

| *Phase I* | *Phase II* | *Phase III* |
|---|---|---|
| AAB Santa Maria | Pendleton Field (330th SG) | AAB Great Falls<br>AAB Rapid City |
| AAB Santa Maria | Biggs Field, El Paso (31st SG) | AAB Pueblo<br>AAB Sioux City |
| Kelly Field | Barksdale Field (307th SG) | AAB Laurel<br>AAB Woodward |
| AAB Syracuse | Fort Dix (28th SG) | Rentschler Field<br>(Langley Area) |
| Daniel Field | AAB Greenville (25th SG) | AAB Florence<br>Myrtle Beach B&G Range |
| Daniel Field | Dale Mabry Field (303rd SG) | AAB Walterboro<br>AAB Charleston |
| Daniel Field | AAB Venice (27th SG) | AAB Sarasota<br>Paige Field, Fort Myers |
| Daniel Field | Avon Park B&G Range (40th SG) | AAB Lakeland<br>AAB Waycross |

Certain details of the new program were still tentative. Indeed, the ASC directive admitted: "It will be noted after a study . . . that not all features of the plan can be carried into effect immediately, but it is desired that the plan be placed into effect in its entirety as rapidly as circumstances will permit." [25] The policy of having parent groups take over the functions of sub-depots, for example, would have resulted in an undue burden of responsibility, and there was a reluctance on the part of the lower commands to combine base administration, technical services, and the training of new units. While Headquarters, AAF, had directed that sub-depots at second and third phase stations were to be drained of personnel after the arrival of parent groups, "and in no instance will this be delayed beyond a thirty day period," [26] there were some organizations, such as the 303rd Service Group at Dale Mabry Field, Florida, which "made no effort to carry out this part of the directive." [27]

Meanwhile, although the Air Service Command was under sharply increased pressure for units to be used in the invasion of Europe, it lacked the priority of the air forces in obtaining personnel and equipment. Just when the build-up of air power in England under the Eighth Air Force was becoming a matter of crucial importance, this command was facing new difficulties in obtaining the wherewithal to meet the proposed shipment dates. The problem was so serious that the Deputy Chief of Staff of the U.S. Army, at that time the Acting Chief, Lieutenant General Joseph T. McNarney, visited Patterson Field and spent a day in conferences with General Frank, General Adler, and the section chiefs of the Personnel and Training Division, discussing the problems and needs of air service units.[28] As a result of these conferences, a brief memorandum, subsequently known as "The McNarney Directive," [*] was dispatched to the Commanding General, Army Air Forces. The larger significance of the McNarney Directive will be dealt with in a later section; in this discussion of the ASC training structure it is sufficient to note that if the question of priorities within the AAF could not be solved by other means, it was directed that "the activation of additional Combat units [would] be deferred until the structure of the Army Air Forces [was] balanced." [29]

As an item of procedure General McNarney suggested that the practice of requiring a unit to be activated "before personnel requirements are recognized" was wasteful and should be discontinued.[30] Also he pro-

[*] There have, of course, been many "McNarney Directives." General McNarney was a trouble-shooter who investigated many situations where the programs of the various branches were at variance.

posed that tactical service units be assigned to air force stations for final phase training, and that they be permitted to perform all functions expected of such units in the theaters. In any case it was vital, he wrote, "to eliminate the current deficit in Tactical Service Units and . . . to bring into balance the Army Air Forces programs for Tactical Service Units and Combat units." [31]

The first indication of the effect of the McNarney Directive was the publication on 25 August 1943 of General Adler's "A Comprehension of the Training Program," [32] which contained a standing operating procedure for the staff units of Headquarters, ASC—a much needed innovation. It was drawn up in the form of a letter to the Chief of the Training and Operations Section, mimeographed, and distributed to key officers at Patterson Field, in Washington, and in the lower echelons.[33] The operation of the McNarney Directive was also apparent in the quickened policy of Headquarters, AAF, in demanding combat-service cooperation from the numbered air forces.[34] Command over the combined training bases, it was stated, would remain with the air force concerned, and Service Groups would function under the operational control of the base commanders. But the administration and training of Service Groups would be the responsibility of the Air Service Command. Both the air forces and the Air Service Command would be informed of contemplated changes in over-all objectives, so that they could alter their plans simultaneously, to preclude confusion or delay in supply.[35]

The operation of this program, however, was interrupted by the announcement in September 1943 of the Bradley Plan, the War Department's proposal for manning and equipping the Eighth and Ninth Air Forces.* This unexpected development, which was highly secret at the time, complicated all planning, for it resulted in the deactivation of many ASC units and the shipment of the personnel of these units to the British Isles as individuals, or casuals. Further training of individuals and the organization of new units was to be accomplished overseas. On the Air Service Command the effect of this remarkable decision was definitely to deflate the domestic training structure, and to cut down the need for service units trained in the continental limits. Henceforth it was necessary to produce but one Service Group per month and one Air Depot Group every other month.

* For further details see section on "The Bradley Plan."

Late in September the Air Service Command was instructed to devise a new, coordinated training plan to implement both the McNarney Directive and the Bradley Plan.[36] In response to this order "A Plan to Stabilize Air Service Command Military Personnel and Training Requirements" was submitted to Headquarters, AAF, by Brigadier General E. E. Adler on 16 October 1943.[37] The conditions existing at the time this plan was submitted were described by General Adler, as follows:

It will be recalled that when the Air Service Command shipped approximately 23,000 individuals under the Bradley Plan, the activated units including those in the emergency pool * were drawn upon to furnish the needed personnel. Following this withdrawal, orders were issued by Headquarters, Army Air Forces, to freeze all remaining personnel in place, pending further developments. As a result, the deflated activated units are now located at many stations throughout the United States and are unable to proceed on any organized unit or combined training program because of the unbalances of personnel. This condition is wasteful of manpower and must be corrected.

In the training structure, now in effect, which is engaged in preparing the committed units on a six months' program for overseas, there are assigned eight (8) Parent Service Groups and sixteen (16) authorized but not fully manned Service Squadrons (Training). As the committed units depart for overseas the need for the Parent Groups and Service Squadrons (Training) as *training monitors* ceases, but their need as a *Service Unit* does not necessarily cease inasmuch as many of these units are located at stations where civilianized Sub-depots cannot be set up. Also one of the Parent Groups is now in the maneuver area of Tennessee serving the Air Forces in that Maneuver. It is apparent that the Air Forces and other Commands are vitally interested in a stabilization of the Service Units.[38]

The new plan was to unfreeze the personnel in activated but depleted units, and set up a changed training structure to produce the recently authorized requirements of one Service Group per month and one Air Depot Group every other month. This program required adjustments in the bulk allotment of the ASC for the manning of training structure units, and in the troop basis of the AAF for the manning of trainee

* An emergency pool consisting of "2 each SG/assoc serv, 2 ADG/assoc serv, 2 Dep Repair Sqns (sep), and 2 Dep Supply Sqns (sep)" was established in July 1943 to obtain priority for personnel and equipment, in accordance with WD AG 320.2 (1-2-43) OB-S-C-M, "Organization Training and Equipment of Units for Overseas Service," 5 Jan 1943, cited in Daily Diary, Air Services Div (Supply & Services), AC/AS, MM&D, 30 July 1943.

units. The flow of personnel to the stations of activation and/or first phase training was to be regulated "so that not less than 85% of the required personnel, basically trained and graduates of the Technical Schools, would be on hand during the first month of activation as required by War Department orders." [39] To facilitate these ends it was requested that Headquarters, AAF, reveal its known requirements for service units in addition to those on the Six Months Commitment List, and specify anticipated readiness dates. For example, "if additional Service Units were to be provided for CBI or to support the B-29 program or to be assigned to maneuvers or other special projects," the Air Service Command wanted to be informed before any of the existing units were inactivated.[40]

Before this plan could be approved in its entirety, basic decisions were required from several sources in Washington, including the Air Staff, the War Department (OPD and G-3), and the Combined Chiefs of Staff.[41] Reinforced by the McNarney Directive, however, the Air Service Command was able to win acceptance for all major points in its program. The personnel freeze was lifted in November,[42] and the distribution of personnel from the activated but not committed units to the elements of the new training structure was immediately begun, on the basis of current information from Headquarters, AAF.[43] The training structure at this point was to be composed of four Parent Service Groups and one Parent Air Depot Group, with the miscellaneous units of the other arms and services. Air depot training stations were to be eliminated, and the number of Service Squadrons (Training) was to be cut down from sixteen to four. It is significant that the training structure was stabilized only after the bulk shipment of troops to the United Kingdom. Manning tables based on the revised estimates of the Air Service Command included the units listed in Table 10.[44]

On 1 January 1944 AAF sub-depots, formerly under the Air Service Command, were reassigned to the air forces and commands, and third echelon supply and maintenance at domestic installations became the responsibility of those commands.[45] Technical control over the former sub-depots was transferred in August 1944.[46] Tactical Service Groups, however, remained under the Air Service Command for technical control and training until the middle of 1945. Meanwhile, sub-depots were placed under the base commanders, and wherever there were parent

TABLE 10

REDUCED TRAINING STRUCTURE

| *Units* | *Officers* | *Warrant Officers* | *Enlisted Men* |
|---|---|---|---|
| 4 Service Groups (Parent) | 256 | 16 | 4,452 |
| 1 Air Depot Group (Parent) | 53 | 3 | 1,021 |
| 4 Service Sqns (Training) | 48 | 0 | 1,120 |
| 1 QM Rescue Boat OTU Det. | 19 | 34 | 102 |
| 1 MP Training Detachment | 30 | 0 | 131 |
| Air Service Command School Det. | 50 | 0 | 206 |
| Medical Training Det. | 10 | 0 | 50 |
| 2 Service Gp Signal Trng Det. | 20 | 0 | 98 |
| 1 Depot Gp Signal Trng Det. | 13 | 0 | 75 |
| 1 Chemical Company, Air Operations | 4 | 0 | 130 |
| 1 Chemical Depot Co, Avn | 4 | 0 | 78 |
| 1 Chemical Maintenance Co. | 4 | 0 | 119 |
| Radar VHF Installation & Maint. | 3 | 0 | 17 |
| Chemical Training overhead | 11 | 0 | 38 |
| | 525 | 53 | 7,637 |

groups, the maintenance activities of the sub-depots were reduced to functions of a custodial nature.[47] In getting this program started Service Centers at operational training unit bases were to function in accordance with AAF Regulation No. 65–1, and also in accordance with a new standing operating procedure issued by Headquarters, ASC.[48]

The training structure of the Air Service Command in 1944 was to be reduced 50 percent, and training loads were to be cut down accordingly. As it turned out, however, there were many unforeseen requests, including: one Air Depot Group and two Service Groups (Special) per month for the very heavy bombardment program, plus six Floating Air Depots, and many other units, urgently required for the Pacific.[49] When these programs were introduced, new training units had to be set up in the Oklahoma City and Mobile ASC's. Also, the training structure at the beginning of the year included the Central Test Pilots School at San Antonio, Texas, and three Personnel Replacement Depots at Warner Robins, Georgia, San Antonio, Texas, and Sacramento, California. Broken down by areas, the national training structure was organized as set forth in Table 11.[50]

TABLE 11

TRAINING STRUCTURE, BY AREAS

WARNER ROBINS AIR SERVICE COMMAND
27th Service Group
40th Service Group
410th Service Sq (Tng)
AAF Med Serv Tng School *
ASC School Det No. 1
Hq & School Hq, Sig Co, Serv Gp Tng Ctr No. 1
770th Cml Dep Co (Avn)
703rd Cml Maint Co (Avn)
Cml Tng Det (Avn) No. 1

SAN BERNARDINO AIR SERVICE COMMAND
307th Service Group

OKLAHOMA CITY AIR SERVICE COMMAND
11th Service Group (now at Pendleton)

SAN ANTONIO AIR SERVICE COMMAND
50th Air Depot Group
421st Serv Sq (Tng)
422nd Serv Sq (Tng)
Hq & Hq Sq, 1st MPTC (Avn)
Hq & School Hq, Sig Co Dep (Avn) Tng Ctr No. 3

SACRAMENTO AIR SERVICE COMMAND
416th Serv Sq (Tng)
Hq & School Hq, Sig Co SGTC No. 2

MOBILE AIR SERVICE COMMAND
835th Cml Co A.O. (M&H)

## PERSONNEL ADMINISTRATION

In as diversified an activity as the military training program of the Air Service Command it was not easy to maintain accurate and current information on all operating agencies. Units and organizations were widely scattered, and were transferred so frequently that it was almost impossible to keep track of them. In fact, during 1942 there was not a single office at Headquarters, ASC, that had the complete picture of any one aspect of the program. Information available in the Supply or Maintenance Divisions, for example, was never in complete agreement with that of the Military Personnel Section, and the figures in Military Personnel were as likely as not to disagree with those in Military Training.[51] Thus, when the Control Office was established by General Frank

* To be redesignated "ASC Medical Det No. 1."

in the reorganization of December 1942,[52] an independent collection of data might well have been developed to confuse the situation still further. It was obvious, however, that greater efforts would have to be made to obtain one set of accurate, coordinated figures.

The Control Office attempted to obtain unimpeachable information on the ASC and AAF training programs: information on combat readiness, strength, housing, aircraft, the location of maintenance and supply facilities, and so forth. Some of this information was obtainable from the statistical reports already submitted. Complete data on strength could be gleaned from AAF Forms 127 and 128; information on housing could be obtained from similar reports. But a description of the status of newly activated units, and the progress of these units in training, could only be obtained by considerable research into the activation and movement orders of The Adjutant General, and the numerous training progress reports submitted to higher headquarters by the units themselves.

The logical agency at Headquarters, ASC, from which to expect information on the training progress of military units was the Training Section of the newly organized Personnel and Training Division. For many months the Control Officer endeavored to obtain information from this section in summary or in chart form for purposes of display and planning, but prior to April 1943 these efforts "mostly came to a dead end." [53] The impediment was that the figures of that office were inconveniently tabulated, and even after they had been rearranged and figured out, they were not at all complete. In March, for example, the Control Room Officer protested that "by putting together the SC-5A and the SC-5B report on other arms and services and throwing the component units into their parent groups, it is found that of the three Air Depot Groups reported, one is ready to perform and two are not. In the case of Service Groups . . . fifteen are not ready, when the component units are included, many of which are not reported on, although the parent groups have been asked to report." [54]

When requests were made by the Control Office for separate, detailed reports on combat readiness, they were ignored for weeks at a time. Such procrastination on the part of the Training Section was ended finally on 23 April 1943 by a brief note to the effect that "this information (the most complete tabulation yet requested) is desired for General Frank *today*." [55] As late as 10 September, however, General Frank reported to the Division Chiefs that nowhere in the areas had he found "charts clearly showing the training of military personnel," and he directed that such charts be prepared in all area headquarters to indicate the number of personnel received, the dates of their arrival, and whether they had received basic or technical training, or no training at all. Such charts, he declared, should "be clear and not have too much on them." [56]

In May 1943 records formerly maintained by The Adjutant General's Office were consolidated with the standard statistical personnel reports of the 15th Statistical Control Unit.[57] At that time the 15th Statistical Control Unit, subsequently connected with General Frank's Control Office,[58] assumed responsibility, in collaboration with the Military Personnel Section of the Personnel and Training Division, for obtaining whatever information was needed on tactical units. It was also responsible for the distribution of T/O's and the allotment of grades and ratings for the thirty-four different types of unit in the Air Service Command.

### THE SHORTAGE OF PERSONNEL

The surplus of enlisted personnel which had been evident earlier did not last long in 1943. (See Fig. V.) As authorized strength rose toward two hundred thousand, with pressing commitments in both fixed installations and tactical units, the selective service system, to say nothing of voluntary recruitment, was increasingly inadequate to meet the demand, and the shortage of all types of specialists became persistent.[59] In the case of officers the situation was even more critical, for there had never been a surplus of officers, and OSC graduates with only a few months' military training were generally less versatile in air force jobs than the more experienced officers of the Regular Army.

Of the available officers, many were incompetent for the tasks at hand. "There is nothing wrong with some of these officers," said one theater commander, "they are just good for nothing, and should never have been commissioned in the first place." [60] Qualified commanding officers were particularly rare; occasionally such reports as the following came back from overseas: "I shudder to think what would have happened had this unit been ordered directly to an active combat zone. . . . I talked with the Group Commander. . . . He had no knowledge of how to install the equipment. He might conceivably have placed wood-working and welding equipment in the same building." [61] So acute was the shortage of officers throughout the army, however, that little was done to reclassify incompetents in this command or else-

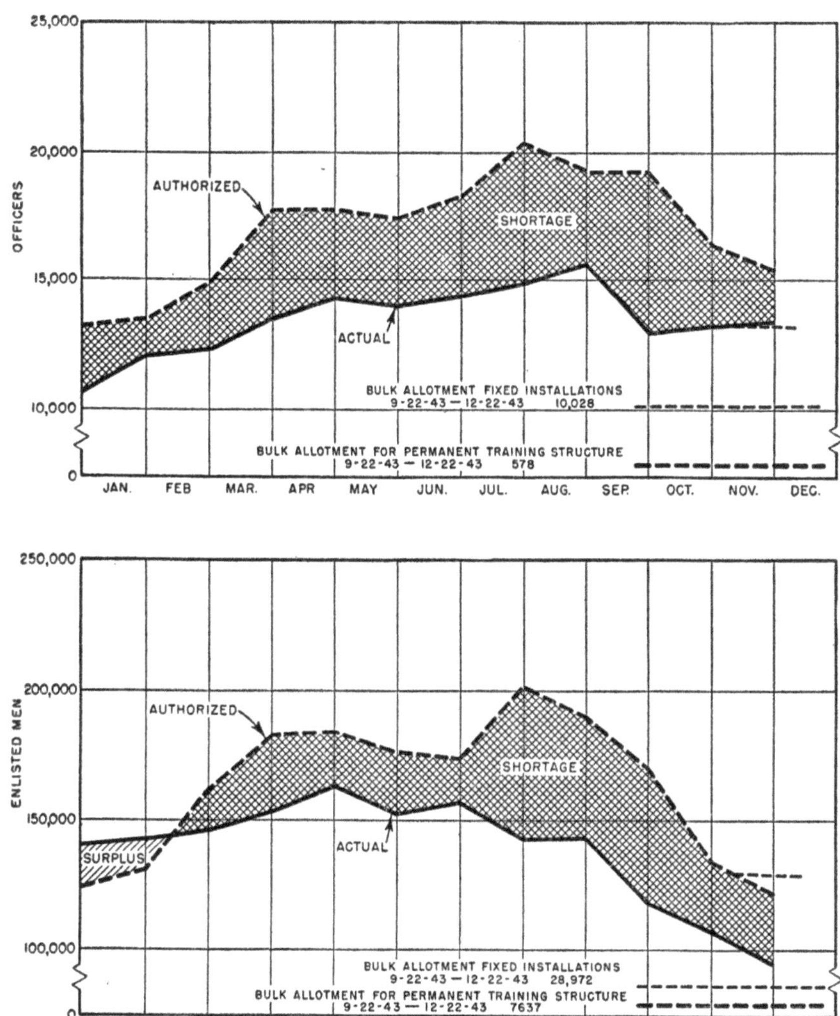


FIGURE V

ASC MILITARY PERSONNEL, ACTUAL VS. AUTHORIZED, 1943
Control Room chart, Hq ASC

where.[62] In the month of May 1943, for example, General Marshall himself reported that of five hundred thousand officers in the army only four were eliminated for inefficiency.[63]

It was felt by some, however, that incompetency was not entirely chargeable to civilians in uniform, but was aggravated by the speed with which Regular Army personnel were promoted. At a meeting of the Division Chiefs of the Air Service Command in November it was suggested as a matter of policy for Regular Army Lieutenant Colonels that "as each case comes up, upon the completion of the necessary time, they shall automatically be sent in for promotion, thereby preserving their relative standing." [64] The meeting concurred in this policy—it was not even necessary for the superior officers of these men to recommend them. Professional soldiers whose only experience had been military were rushed up through the ranks, and entrusted with heavy responsibilities in a highly technical, industrial business.[65]

To relieve this shortage several attempts were made to recruit qualified volunteers from civilian life and from the ranks. The outstanding campaign of this sort in 1943 was an effort to obtain 550 "maintenance and supply technicians" (200 of the former and 350 of the latter). Quotas of 50 were established for each of the areas, and boards of 3 to 5 officers were set up to consider the qualifications of the applicants,[66] but only "mediocre results" were obtained,[67] and by 15 October 1943, after 150 officers had been procured by this means, direct commissions were terminated.[68] Thereafter the only commissioned personnel available to this command were OTS, OCS, and Flying Training School graduates, or returnees from overseas.[69]

Meanwhile, the shortage of ASC personnel was recognized by the War Department, and unusual steps were taken to alleviate it. In June officers from several ASF Service Commands were released for transfer to ASC, but again the number "was far short of the 550 expected by at least the 15th of the month." [70] Enlisted men were also transferred to this command. In July, for example, 1,406 Air Corps enlisted men and 648 men from the other arms and services were transferred from the Technical Training Command, and comparable numbers were transferred from the Flying Training Command and the Army Service Forces. But the supply was still insufficient; "while helpful," the Weekly Activity Report stated, "this number of individuals did not materially aid toward filling the shortages in specialties of approximately 29,000 EM and 2,700 officers."[71]

Emergency calls for special troops, such as maintenance and supply personnel for the Burtonwood Project at San Antonio, added to the predicament of the Air Service Command. This project called for the furnishing within a period of two months[72] of an entire complement of technically trained military personnel to operate a Base Air Depot under the Eighth Air Force in England.[73] On instructions from The Adjutant General,[74] each of the eleven areas was given a quota of voluntary inductees. These were to be sent to San Antonio for specialized training, and shipped overseas during August and September in two large casual shipments of approximately 2,000 each.[75] With the exception of a few technicians needed for newly developed equipment, these voluntary shipments were successfully filled.[76]

Under ordinary circumstances enlisted men were allotted to the Air Service Command in bulk, and thence reassigned. For a time, incoming personnel were directed to the areas in accordance with a "blind prorating," but in May 1943 it was determined that assignment should be based on the most urgent requirements.[77] Ideally, personnel were to be sent to first phase training stations; more often they were shipped as last-minute fillers to third phase stations.[78] It was suggested in June, as an emergency measure, that unfilled units be deactivated until they were needed,[79] but this policy was not inaugurated until after the announcement of the Bradley Plan. The administration of incomplete units, in some cases one-man units, was continued despite all inconveniences.[80]

The outcome was that service organizations had to begin unit training before they were fully or competently manned. There was also a constant depletion of trainee units well along in group training to fill other units more immediately committed. New units were activated without personnel in order to obtain priority for the allotment of troops. Sometimes the troops were never allotted; at other times they were not allotted until much later.[81] Although cadres of officers and enlisted men should have been on hand long before the beginning of second phase or unit training, the personnel were not in fact available and conventional preliminaries were often omitted. Consequently, filler personnel in large numbers—basics from Reception Centers—were sometimes assigned to service organizations after the receipt of warning and movement orders.[82]

Of the many factors related to personnel administration which "contributed to delay and confusion in the training of Service Groups," the following were considered the most important:

(a) Shortage of officers, and failure to secure same in advance of unit activations;
(b) Over-all shortage of enlisted men, and shortage in specific trades;
(c) Loss of enlisted men in the final stages of training to OCS's (often replaced by basics);
(d) Loss of limited service enlisted men in the final stages of training (also often replaced by basics);
(e) Assignment of individuals unqualified for service with ASC because of a lack of education or intelligence;
(f) Completion of many aspects of training before the assignment of personnel;
(g) Delay in moving units to the port after they were ready to go, thereby checking the normal flow;
(h) Lack of adequate classification centers.[83]

The same faults were found by inspectors in group after group.[84] Over a long period the reputation of the Air Service Command became so poor with higher authority that General Adler was obliged to report in the spring of 1943 that "it had become *normal* for the Chief of Staff to receive derogatory reports on the combat readiness of ASC units." [85] On several occasions letters in words-of-one-syllable were received from General Arnold with such blunt remarks as the following: "It has been brought to my attention that not once, but repeatedly, service units . . . have not passed the inspectors . . . due in some cases to improper training . . . and in still others to the lack of properly disciplined personnel. We have been engaged in shipping units overseas for such a period of time now that such delays as this are inexcusable." [86] Under the circumstances the best the leaders of the Air Service Command could do was to reply vigorously to each criticism with a complete statement of the difficulties under which they were laboring, and the obstacles with which they were faced.

It was pointed out, for example, that it was necessary to schedule service units for overseas shipment long before the command had received, or knew when it would receive, either the personnel or the equipment, and that higher headquarters had frequently requested for shipment more units than the Air Service Command had been willing to promise. Since the Air Service Command was anxious that the theater commanders get everything humanly possible, it had consented to these increases—with the understanding that the units would be ready by the time they were shipped, but not necessarily a month in advance, when they received Preparation for Overseas Movement (POM) inspections.[87]

If the Air Service Command were to have its units ready for POM inspections, the command requested either that it be granted an additional month for training them,[88] or that it be permitted to waive some of the requirements for military training.[89] In the latter connection, General Adler pointed out in a conference with the Chief of Air Staff, General Stratemeyer, that service units would generally be stationed in rear areas, and so have time to complete their military training after arriving in the theaters. Technical training was the only critical item, and this would be accomplished in the zone of the interior before the units were shipped.[90]

Actually such requests were made not so much in the hope that they would be granted as to impress higher headquarters with the importance of furnishing additional personnel, and to register each difficulty "for the record." In point of fact, the Air Service Command had become quite fatalistic in its attitude toward the receipt of personnel. "Unfortunate instances," said General Adler, "must be expected as a natural incident to the law of averages." [91]

On rare occasions units went through training on schedule all the way down the line. Such was the case of the 39th Air Depot Group, which was trained in record time at San Bernardino, and later won the highest praise of the Commanding General of the Alaskan Defense Command, Lieutenant General Simon Bolivar Buckner. One of the main factors in the success of this organization, wrote Brigadier General Lucas V. Beau, Commanding General of San Bernardino ASC, was "the physical presence of the necessary personnel at the time of activation of the unit." [92] A letter like this from the field confirmed the statements of General Adler and others at Headquarters, ASC, that many of the deficiencies of service units could be traced to the shortage of personnel.

Similar opinions were forthcoming from the Commanding Generals at Sacramento, San Antonio, and Warner Robins, who requested in August that they be permitted to have at least "70% in personnel" before being ordered to activate new units.[93] In reply they were informed that, as a result of General McNarney's visit, the Air Service Command had been authorized "85% of T/O strength by SSN" prior to activation of a unit.[94] But they were not told, for it was not yet known, that the personnel were still not available.

### CENTRALIZED PERSONNEL REPLACEMENT DEPOTS

In 1942 certain designated units in each of the areas served as replacement centers for the classification, preliminary training, and assignment of ASC personnel. These units operated under fixed T/O's with authorized overstrengths, often sizable ones, but practically always the actual overstrength was far greater than that authorized. Furthermore, many units were used as pools for unassigned personnel without any authorization at all,[95] except that such authorization had been requested. As the Air Inspector, Brigadier General Junius W. Jones put it: "It is admitted that there have grown up unauthorized activities pending the approval of changes in T/O's (which changes take considerable time). Having adopted such unauthorized activities, the changes in T/O's may or may not be further urged, but it is regarded by officers of the ASC that the suggested changes are on file for the sake of the record." [96]

Justification for these replacement pools was not hard to find; experience had shown that incoming personnel received by the Air Service Command from Reception Centers, and even from the Technical Training Command, arrived "in all sorts of conditions—poorly classified, not paid for long periods, with no records, bad teeth, no basic training, etc." [97] If omissions and mistakes of this sort were to be caught, the best time was immediately after the personnel had been assigned. Also, it was highly desirable to have a proper T/BA for the administration and training of all troops before sending them to tactical units, such equipment as typewriters, office supplies, and various items of organizational equipment being critical and unobtainable without priority.[98]

Efficiency, therefore, seemed to demand centralized control over a limited number of replacement centers (which might be called wings, battalions, or depots) rather than a scattering of separate units throughout every area in the country. For this reason the Air Service Command repeatedly requested authority for the activation of three replacement centers, to be located at Daniel Field, Georgia; Kelly Field, Texas; and

Santa Maria, California.[99] For several months, however, this authority "had not been favorably considered," partly because of the shortage of personnel and partly to avoid conflict with the pending system of bulk allotments.[100]

The creation of unauthorized (or "bootleg") activities was the start of a vicious cycle with ramifications affecting every aspect of the training program. Personnel in excess of recognized strength were reported to headquarters on AAF Form 127 [101] as available for reassignment, and requisitions for personnel based on shortages were automatic.[102] Thus, when Headquarters, AAF, insisted that overages be eliminated to fill shortages elsewhere, compliance in some cases was possible only by an evasion of the classification regulations. At that point, poorly classified individuals, without interest, aptitude, or technical background in their specialties, were transferred to tactical organizations with the same classifications as those with which they had arrived at replacement pools; or else true talent was "buried" by conscious misclassification to keep it on hand for more careful disposition later.[103] When criticized for these practices, ASC leaders replied that after an initial allotment had been made, the shifting of personnel should be left entirely to this command.

Without doubt, AAF Form 127 was a poor yardstick for the shortages and overages of training stations and replacement centers which were not based on authorized T/O's.[104] It was assumed that the current T/O's were correct for the situation at hand, but until exact manning tables and bulk allotments were instituted, this was not the case. For instance, on 15 August 1943 the Air Inspector reported that the 127 report from Daniel Field, Georgia, indicated a shortage of three "crash boat operators" and an overage of thirteen "athletic directors," but that the addition or removal of personnel on the basis of these figures would be wasteful, for there was not enough flying at Daniel Field to require the crash boat operators, while the large number of men received for replacement made every available athletic director necessary.[105]

Even in tactical organizations the 127 reports were confusing, for some Service Groups had only one Service Squadron, and some Air Depot Groups had only one Repair or Supply Squadron, instead of the usual number. Thus the 7th and 58th Service Groups and the 32nd Air Depot Group were reported as short in personnel on 28 July 1943 because member units had been ordered overseas, although the remaining units were at full strength.[106] For such reasons, incorrect statements of shortages or of requirements were obtained by substracting the men present from the group's total T/O strength. Thus, in the words of the Air Inspector, when AAF Form 127's were worked up nationally, they "resulted in all of these individual problems adding up into one big mistake." [107]

There were also errors which came from the juggling of figures. The command believed in keeping trainee units as nearly complete, on paper, as possible. Consequently, men at schools, or away on extended periods of temporary duty, were carried in the books by replacement squadrons as overages, and their company positions were listed as filled. For example, on 20 August 1943 the Warner Robins area had 856 persons listed for the Headquarters Squadron of Daniel Field alone—and nowhere on the record was it apparent that these men actually were away from the base. Another unfortunate practice was the setting up of unauthorized schools as a means of building needed reserves in certain critical specialties.[108]

The chief incongruity of the whole situation, however, was that each area had shortages of specialized skills which could never be filled from the local pools, because neither the area headquarters nor Headquarters, ASC, could control the filling of the pools.[109] Personnel were "dumped" on the command by AAF Basic Training Centers, and by the Army Service Forces, with little regard for the appropriateness of their assignment, a state of affairs which caused General Adler to remark: "There is no objection to the Air Service Command being the receiver of mentally deficient or otherwise useless personnel, but it must be established in the minds of the Director of Personnel and other interested Headquarters that such surplus and useless personnel cannot be allotted to tactical organizations of the Air Service Command." [110]

The command sought to relieve malassignment through roving classification teams, which circulated in each area to examine classification cards and interview the men. Theoretically, these teams were detachments of the Military Personnel Sections of the area headquarters, but they were not always posted on the over-all needs of the area, let alone of the Air Service Command.[111] They had a tendency to follow the line of least resistance, and to reclassify on the basis of local overages and shortages.[112] A spot study by the Air Inspector, however, indicated that classification in the Air Service Command was generally better than in

AAF Basic Training Centers, and that assignment to technical schools was connected with intelligent efforts at proper classification.[113]

On 3 September 1943, as has been pointed out, the War Department

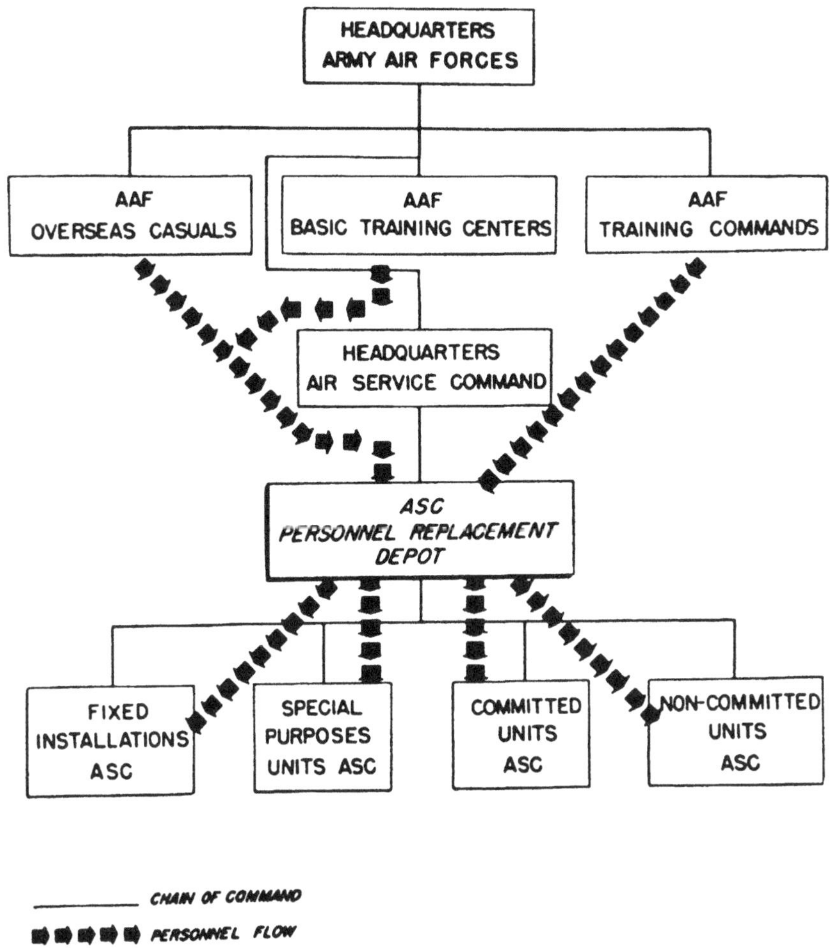


FIGURE VI

ASC PERSONNEL REPLACEMENT DEPOT

finally authorized the activation of three replacement depots under the jurisdiction of the Air Service Command. (See Fig. VI.) ASC Personnel Replacement Depot No. 1 was to be established at Daniel Field, Augusta, Georgia, to service Warner Robins, Middletown, Rome, Atlantic

Overseas, and Fairfield ASC's, and the Miami Air Depot; Replacement Depot No. 2 was to be established at Kelly Field, San Antonio, Texas, for San Antonio, Mobile, and Oklahoma City ASC's; and Replacement Depot No. 3 at Fresno Basic Training Center, Fresno, California, for Sacramento, San Bernardino, Spokane, Ogden, and Pacific Overseas ASC's.[114] Two months later the 15th Replacement Control Depot was established at Daniel Field, Georgia, to operate in conjunction with ASCPRD No. 1 in the handling of all colored replacements of the Air Service Command.[115]

Inasmuch as classification and assignment were to be consolidated in these personnel replacement depots, it was anticipated that the benefits of centralized control would immediately become apparent. Not least of these benefits would be the filling of Specialty Serial Number (SSN) shortages with surplus and unauthorized SSN's through lateral reclassification and assignment to specialized schools. By these means, for example, SSN 040 (auto body repairman) could be transformed into SSN 555 (sheet metal worker) in which there was a critical shortage.[116] Another advantage would be a greater facility in handling returnees and other assignees from Redistribution Centers; those from the West would be shipped directly to Fresno, California, and those from the East to Warner Robins, Georgia. Formerly all returnees allotted to the Air Service Command had been sent to Headquarters, ASC, Patterson Field, Ohio.[117] Thus, an improvement in personnel administration would facilitate the management of the training structure and the flow of the trainee units through their various phases. (See Fig. VII.)

Basic military training was to be emphasized at replacement depots,[118] although many individuals had been through it before, because such training would develop discipline and "keep the men out of trouble." [119] It was to supplement the military training at AAF Basic Training Centers, and to qualify all personnel for general duty in a period of fifteen days.[120] By 1 November 1943 a standing operating procedure for training, classification, and assignment was under preparation at San Antonio, and shortly thereafter it was operative throughout the Command.[121]

### THE MC NARNEY DIRECTIVE

Meanwhile, on 11 August 1943 Lieutenant General Joseph T. McNarney, Deputy Chief of Staff of the U.S. Army, descended upon the

Air Service Command without warning to inspect "all phases of the military training set-up." Evidently the reputation of AAF service organizations had become so bad that the Army's No. 1 trouble-shooter had undertaken to improve it. Prior to that time, on a number of occa-

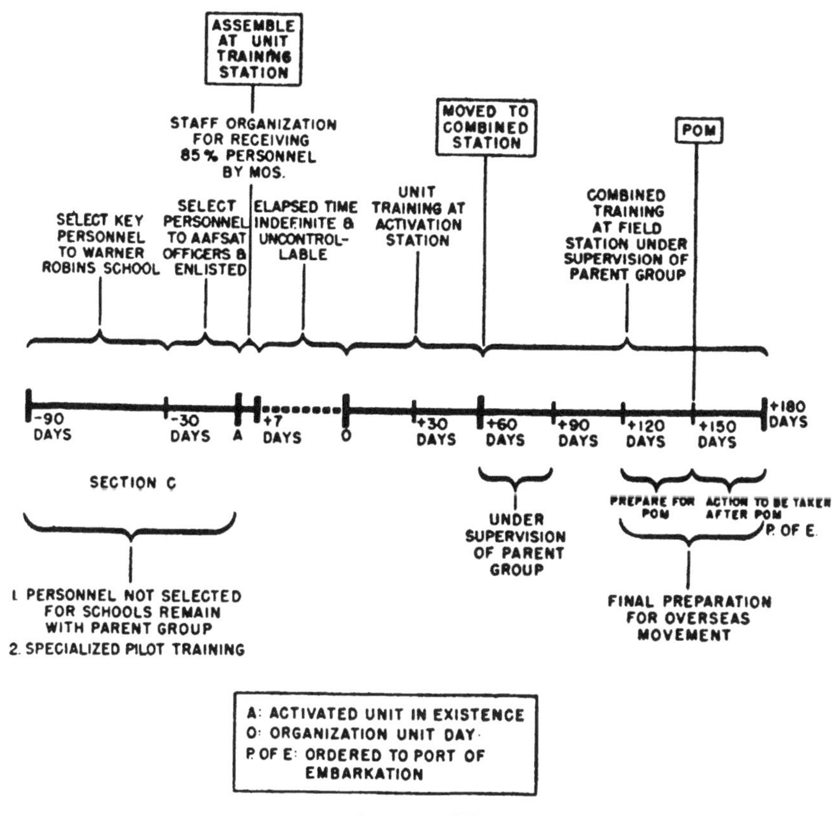


Figure VII

Schedule for Organizational Development

sions, responsible generals in Washington had been called to General McNarney's office to explain the reasons for unsatisfactory training status reports on air service units; [122] now, apparently, the Deputy Chief of Staff had come to see for himself.

No one at Headquarters, ASC, had been aware of the General's intentions until the evening before he arrived, and during the two days he spent at this headquarters he proved to be an energetic visitor.[123] In addition to lengthy conferences with General Frank and General Adler, he visited the Control Room, inspected the charts in the office of the Chief of Training and Operations, talked to enlisted men in the hangars along the flight line at Patterson Field, and singled out at random for detailed investigation a Quartermaster Truck Company, which was located in the training area for Air Depot Groups.[124] Four days later, at the request of General Arnold, a meeting of the Air Staff was convened in Washington by General Giles to allow the leaders of the Air Service Command to explain just what went wrong while General McNarney was at Patterson Field.[125]

Among other things, the young officer in command of the Quartermaster Truck Company had told General McNarney that although he had had no training himself, he was expected to give basic military training to his men without help; that no one had been down to instruct him. General Giles reported that General McNarney had felt that with two higher headquarters on the same base there should have been more supervision of training than was evidenced in this case. General Adler's rejoinder in connection with this incident was:

> I want to add that the CO of this organization was apparently a little taken aback by the presence of the Chief of Staff. He has changed his story somewhat, and now says he has had some instruction. A statement is being taken from him to this effect, which will be in my office. I hadn't intended to bring this up, but just mention it at this time.[126]

General McNarney's inspection, however, had not been limited to minor incidents of this nature. When he flew to Dayton, he was aware of the (still secret) Bradley Plan for the build-up of the Eighth and Ninth Air Forces in the British Isles, and knew that the staff of G-3 of the War Department was working out a way in which this plan could be implemented. He did not know what the solution would be; the great decision (for which he himself would be responsible) had not yet been made.[127] His primary concern on this visit was to speed the training of service units. He wanted to learn at first hand the difficulties of the Air Service Command in its relationships with other commands, and wherever possible to reconcile conflicting interests. Above all, he wanted to dramatize the importance of the Air Service Command's training program.

In the course of his conversations at Patterson Field General Mc-

Narney got to the heart of several vexing problems; the significance of his visit lay in the success with which he crystallized these problems in his own mind and in the minds of those with whom he talked. The training of service units, he said, was "out of phase" with the training of combat units. It had not received sufficient priority in terms of personnel and equipment. In the future, it should. If the Army Air Forces was unable to balance its program in any other way, it should defer the activation of combat groups until Service Groups and Air Depot Groups had been allowed to catch up.[128] Originally, General McNarney stated, the Air Service Command with its Eighty Air Depot Group Program, had been far ahead of the air forces, but recently it had been unable to keep pace. The reasons for this dangerous condition were the lack of coordination between the War Department and the Army Air Forces, and inadequate over-all planning. That the justice of these observations was appreciated by the members of the Air Staff was indicated by the following exchange:

General Giles: ". . . Among other things Hq, AAF has failed in over-all planning to determine how many service units are needed for the AAF."

General Whitten: "We do have a plan covering the number of service units."

General Frank: "That's true, but the plan has been changed too often."

General Whitten: "That is right. But there have been additional requirements—for the Air Transport Command, for the militarization of sub-depots, and so forth."

General Giles: "We must have a stable program. . . ." [129]

The basis for such remarks was not hard to find. Under the 273 Group Program, as planned in September 1942, the service needs of the entire Army Air Forces were to have been met by the activation of 81 Air Depot Groups and 160 Service Groups, roughly computed on the basis of one Service Group for two operational groups and one Air Depot Group for two Service Groups.[130] Early in 1943, additional, unanticipated requirements had arisen for the routes of the Air Transport Command in Africa, India, and China [131] and for the militarization of the domestic sub-depots. Although the Air Service Command had more civilian than military personnel, it was impossible to employ any large number of civilians in sub-depots at such relatively uninhabited places as Muroc Lake, California, and Tonopah, Nevada, because civilians refused to live and work in deserted areas.* Thus the service needs of the AAF had become greater than expected, and the demand for military personnel in the Air Service Command had been correspondingly increased.

Meanwhile, during 1943, it had become apparent that there would be an element of geographical stability in the war. The Air Staff realized that "as long as the war goes on, England is going to be a base; . . . that India is a base; that Australia is a base; and . . . that Africa will be a base until or unless [the war] is moved over on the other side of the Mediterranean." [132] Even so, it was difficult to determine the precise relationship for service and operational units, applicable to all theaters, and the rough computation originally used had to remain the guide for all but exceptional cases. The relationship between service units and operational units had therefore remained vague and indefinite, largely because of the uncertainties of the war itself. But an important contributing factor, as General McNarney was free to admit, had been the slowness of the War Department in coordinating the requirements of the theater commanders (for combat units), the Air Transport Command (for its overseas routes), and the Air Service Command (for its militarized sub-depots).

So far as the AAF was concerned, one of the first changes arising from General McNarney's visit was a reorganization which transferred the supervision of all training activities to one staff element at Headquarters, AAF, in Washington, D.C. In this reorganization, responsibility for the training of air service units was transferred from the Assistant Chief, Air Staff, Materiel, Maintenance, and Distribution to the Assistant Chief, Air Staff, Training,† who also supervised the training of operational units.[133] The purpose of this change was to enable Headquarters, AAF,

* In 1942 and 1943 various proposals were made for the militarization of ASC activities. (See DAR, Hq ASC, 17 June and 13 Oct 1943, and WAR, Hq ASC, 12 March and 5 Nov 1943.) Brig Gen Lester T. Miller went so far as to recommend the militarization of all ASC installations in the U.S., except the depots and the Field Service Section. (See *ibid.*, 25 Oct 1942.) This extreme proposal was never carried out; by 1 Jan 1944, when the AAF sub-depots were transferred to the commands and air forces, 23 of the 238 sub-depots had been militarized, primarily with personnel not qualified for overseas duty. (See "The Air Service Command, An Administrative History, 1921–1944," p. 79; also WAR, Hq ASC, 17 Dec and 28 Dec 1943, in TSHIS-2 files.)

† Formerly responsibility for the training of air service units had been exercised by the Directorate of Base Services, which had been absorbed by AC/AS, MM&D on 29 March 1943.

to bring the training of service units "into phase" with that of other units.[134] At first, its success was not unmixed, for the two staff agencies continued their "disputes as to the functions to be performed by each" for a period of several months.[135] Later the training program functioned more smoothly.

In addition to this reorganization, which was something of a by-product of General McNarney's visit, "The McNarney Directive" provided, as has been stated above, that service units were to be granted a priority equal to that of combat units; [136] it also made a number of administrative changes. Organizations for which there was not a minimum of 85 percent personnel strength were not to be activated. Air Base Security Battalions were to be eliminated,* and additional personnel were to be procured as soon as possible from the Army Ground Forces and the Army Service Forces.† Appropriate field training was to be emphasized throughout, including such simulated combat conditions as bivouacs, movements, concealment, and camouflage. Finally, Service Groups were to be assigned to air force stations, at which subdepots would be militarized, for final phase operational training with combat groups. These were General McNarney's provisions for a stabilized training program. The problems had been clarified; it now remained to be seen if the directives of the Deputy Chief of Staff could be complied with under the impact of future developments, especially the Bradley Plan.

### THE BRADLEY PLAN

Key individuals in the Personnel and Training Division, ASC, had often been warned to "expect the unexpected," [137] but they were hardly prepared for the Bradley Plan. From the beginning of the war emphasis had been placed on the training of units; the idea that partially trained, or untrained, individuals should be shipped overseas by the thousands, as casuals, seemed a throwback to the methods of World War I. It struck without warning, but perhaps for that reason the whole undertaking was carried through with unusual success.

The plan itself was the work of Major General Follett Bradley, who

* The inactivation of Air Base Security Battalions released 93 officers, 2 warrant officers, and 5,275 enlisted men (colored), and 38 officers, 4 warrant officers, and 898 enlisted men (white). (See WAR, Hq ASC, 27 Aug 1943.)

† Two hundred officers were procured (⅓ in Sept, ⅓ in Oct, and ⅓ in Nov) from the AGF and the ASF, 85% of company grade. (See WAR, Hq ASC, 27 Aug 1943.)

had been sent to the European Theater to report on the combat effectiveness of the Eighth Air Force, and to recommend means of manning and equipping both the Eighth and the Ninth.[138] The purpose of the Bradley Plan * was to lay a foundation for air power, based on the United Kingdom, powerful enough to soften up German resistance and facilitate the invasion of the continent. D-Day was set for the first of May 1944, or thereabouts, though German air defenses were unremittingly effective; it was apparent that heavy reinforcements would be required.

Increased demands on the AAF were caused by the need for greater speed in the building of this force, and by higher attrition rates in the United Kingdom than those anticipated in earlier estimates.[139] The idea of shipping individuals as casuals was not part of the original plan, but implementation of the Bradley Plan involved more than had been contemplated in the 273 Group Program, and verbal inquiries had indicated that too few additional units were available.[140] In fact, it was already difficult to meet even the routine requirements. All elements of the Army Air Forces were forced to ship casuals; the Air Service Command soon found itself continuing irregular practices which had been "forbidden" by the McNarney Directive, such as the activation of new units "before personnel requirements [had been] recognized." [141]

On hand in England, or en route by 7 September 1943, were 167,806 officers and enlisted men of the Air Corps and the other arms and services attached thereto. Authorized specialists in units or combat crews scheduled for shipment during the following ten months numbered approximately 180,000, but total Bradley Plan requirements exceeded 500,000.[142] Consequently, 165,000 fillers (in addition to planned AAF requirements) still had to be shipped to the United Kingdom,[143] and the Air Service Command had to contribute its share.

Preliminary studies in Washington had indicated that the Army Air Forces was unable to furnish trained units in the number requested, since neither the combat units nor the service units were available at that time.[144] For this reason, and because of the limitations on AAF shipping space, the War Department proposed to send some of the units intact, and the remainder as individuals, the theater commander

* The Bradley Plan, in spite of its name, was never a single, clearly defined program; it was a series of recommendations which served chiefly as a basis for further planning. Prepared by General Bradley in the spring of 1943, it was adopted by the War Department early in the fall. By that time it had been much altered, and the schedule of shipments continued to change from week to week.

concurring. The idea was that until such time as the Army Air Forces was able to furnish completely trained units, as required by the original plan, it would be necessary to furnish fillers, and to authorize the Commanding General, Eighth Air Force, to activate in the theater whatever units he might need. Such a program necessitated preshipment of equipment and a good deal of guessing as to requirements,* but it was the fastest way to build up the combat strength of the theater.

The Bradley Plan increased the AAF Troop Basis to 2,225,000, including 70,000 in the College Training Program, and required a minimum monthly intake into the Army Air Forces of 45,000 enlisted personnel.[145] In a period of eight months this program would provide 165,000 men for the European Theater, of whom 128,000 would be trained or partly trained in some specialty, and 37,000 would be basically trained only.[146] Several major adjustments were required in personnel allotments for overseas and domestic programs; the one affecting service units most directly was the provision that the Training Command should "reduce the input into technical training to that figure which would produce 20,000 graduates per month, including ASWAAF (white and colored), and stop factory training immediately." [147]

In the case of tactical service units, it was determined that the Bradley Plan should be accomplished in the following manner: service units already committed on the Six Months Commitment List were to be shipped to the United Kingdom as scheduled; personnel left over were to be shipped as casuals; and units activated but not committed were to be inactivated, and the personnel so released were to be used to bring committed units to T/O strength, or else to be shipped separately. Altogether the Bradley Plan called for 40,000 AAF troops per month.[148] The schedule of inactivation applied to all ASC units except: those on the current AAF Six Months Commitment List; those required for use in this country (parent groups, emergency pool units, AAFTAC units, Desert Training Center units, and so on); and those being trained for special purposes (as the 96th Service Group, colored, which would service colored combat groups).[149]

On 10 September 1943 General Adler learned that the first casual shipment of 20,000 ASC troops was to be ready to sail on 25 September. When these instructions were given to him over the telephone, General

* The subject of requirements is treated in greater detail in the section entitled "Organizational Equipment."

Adler protested that the Bradley Plan would be "The Wreck of the Hesperus," and General Whitten in Washington was inclined to agree with him,[150] but the decision could not now be changed. It was arranged that the needed personnel would be dispatched to staging areas immediately, and distributed in balanced categories so that if certain shipments were lost, no critical categories would be completely unavailable.

It soon became apparent that AC/AS, Training, and other agencies at Headquarters, AAF, were desirous of retaining as many ASC units on active status as could be kept with proper justification.[151] Although the necessity for building the theater strength was appreciated by all, there was still the feeling that the chief burden of training should be borne by the zone of the interior. General Adler summed up this point of view at a Theater Supply Conference in November 1943 with strong sarcasm:

> The cry came for the bodies: "Send the bodies over, and as long as they are warm we will take care of the rest; we need the manpower over here." Now I take the liberty of disagreeing with that as a good principle. I think the burden of training should remain within the United States . . . but that is only my personal opinion and has no backing from anybody else. . . . I mention it because I think some of you will be tempted from time to time to be shrieking back: "Just send us the bodies, and we will carry on." Well, we will do it if we are ordered to do it, but I don't like the idea myself.[152]

While General Adler was conferring with Washington, his Assistant Chief initiated a long-distance conference conversation with the eleven area commanders, outlining the unprecedented program and "alerting" them and their staffs for future directives.[153] They were told at this time that the soldiers were to be prepared both physically and administratively, in accordance with the War Department standards in POM,[154] and that especial care was to be taken to avoid a medical bottleneck. Technical personnel were to carry their hand tools with them (in spite of the protest of some of the ports),[155] just as Infantry soldiers carried their weapons, and this provision was to apply to other arms and services technicians as well as to aircraft mechanics.*

Subsequent administrative instructions were expected to be, and

* The Signal Corps, for example, was to contribute 300 officers and 3,000 enlisted men, one-sixth of the total. Although a delay in the shipment of officers was requested, the bulk of this quota was met. (See WAR, Hq ASC, 19 Aug 1943 and 15 Sept 1943.)

were, a great deal more specific. Although the shipments were not all identical, the conditions under which they were conducted were illustrated by the following teletype directive:

> . . . This personnel will be assembled in casual units consisting of 5 officers and 200 enlisted men, including one Medical officer for each 500 enlisted men. All SSN's will be certified by base classification officer. All service records, classification cards, immunization registers, and allied papers will accompany enlisted men. Arms will be issued at port. Officers and enlisted men will be equipped in accordance with TE-21, as amended 10 March 1943, winter clothing only. This headquarters will be notified not later than . . . of the station of each casual unit. Officers and enlisted men will have completed familiarization course on carbine and qualified with any weapon in accordance with POM. Enlisted personnel will be issued one barracks bag, duffle, after arrival at port to replace barracks bag A and B. Barracks bag issued to enlisted men at port staging area will enable transfer of all authorized and individual equipment into duffle bag. Any surplus items and equipment which cannot be packed into duffle bag will be confiscated at port. Retention of one barracks bag for enlisted men to serve as laundry bag is authorized. Stencils will not be available at port to mark duffle bag and will be carried from home station.[156]

The first shipment was by far the largest; it consisted of 17,000 enlisted men and 3,000 officers.[157] Provisional squadrons of 200, combined for command purposes into seventeen key units of approximately 1,200 each, facilitated control of this amorphous crowd, and the entire program was carried out successfully. Subsequent shipments were smaller, usually between 2,000 and 3,000 enlisted men and from 50 to 100 officers.[158] In fact they were so small that for a time there was criticism to the effect that the so-called Bradley Plan was not being adhered to, for a review of the build-up of the Eighth Air Force seemed to indicate that the War Department's desire for 40,000 AAF personnel per month was not being complied with. So far as the Air Service Command was concerned, however, the discrepancy was caused by the fact that the War Department was attempting to extract from this command a total of 95,000, for all purposes, as against an available 56,000. Once the bulk allotment was adjusted, a suitable priority for the United Kingdom (as distinguished from the other theaters, the domestic training plan, and the fixed installations) was reestablished.[159] Also the training program was reduced, as noted earlier, to one Service Group per month and one Air Depot Group every other month.

### Organizational Equipment

In the training of service units the Air Service Command was responsible not only for personnel administration, but also for the supply of organizational equipment. The same administrative trend can be seen here: the gradual development of a unified system, which was suddenly to be altered by the Bradley Plan. As procedures were evolved, the system of supply was changed several times. Originally, organizational equipment was shipped to the stations of activation; at a later period in the war it was shipped to the first phase training stations; and finally to the staging areas, the Ports of Embarkation, or directly to overseas Base Air Depots.[160] The effect of these changes on the training of service units was not inconsiderable, for the possession or the lack of possession of organizational equipment was closely connected with the conduct of technical training. Moreover, this equipment was needed not only for the training of new units, but also for the operation of parent groups. To a large extent the success of the Air Service Command and the numbered air forces in establishing combat-service training programs was dependent upon efficiency of supply.

In 1941, with the introduction of the Organizational Equipment List (OEL) as an automatic requisition for Air Corps supply,* attempts were made to ship organizational equipment to each unit's station of activation. This procedure worked effectively when only a few units were involved, and when the supply sergeants of each squadron were familiar with army procedures, but it was not adequate once the precipitous expansion of the Air Corps had begun. Furthermore, supply agencies were given little time under this system to assemble complete sets of equipment, and often the units were moved from activation stations to other stations before their equipment was dispatched. Complaints were received from all over the country to the effect that shortages remained unfilled for months after the OEL's had been prepared, and that some organizations actually went overseas before their full equipment caught up with them.[161]

Early in 1942 it was determined that organizational equipment should be sent to first phase training stations, instead of to stations of activation, thereby giving supply agencies more time to provide the materiel,

* SOS (ASF) supplies and equipment still had to be requisitioned in the usual manner. OEL's applied only to AAF equipment.

and perhaps eliminating the necessity for forwarding overdue shipments. It was realized that newly activated organizations had no equipment whatever, and the formality of compelling them to prepare their own OEL's was useless. The new system, theoretically, provided equipment with which technical, on-the-job training could be given during the brief unit training period.[162]

When warning orders were received, secret movement numbers were given to each shipment and all available equipment was packed by the unit concerned, marked with the designated movement numbers, and sent to the Port of Embarkation, where last-minute shortages were filled. Warning orders and movement orders, however, were rapid and unpredictable, and teletypes listing shortages were sometimes twenty, thirty, or even forty feet long.* Consequently, in some cases supply was still delayed until after the units had sailed. To provide more time for the processing of shortage lists, a modification was adopted whereby unit commanders were required to conduct formal "showdown inspections" immediately upon receipt of the warning orders, while the units were still at final phase training bases.[163] Shortages revealed by these inspections were to be filled by the base commanders, or by the subdepots or depots from which the units received their supplies. All other requisitions were to be cancelled when these shortage lists were submitted.[164]

Too much responsibility, under this system, was placed on unit personnel, who should not have been expected to be experts in supply. Moreover, it was discovered that many pieces of equipment, initially supplied, were worn out, lost, or destroyed during training.[165] Experience indicated also that materiel shipped to the Ports of Embarkation was often misplaced as a result of the congestion of freight. As Headquarters, ASC, had no record of such shipments, no action could be taken to locate and forward them to their proper destinations. The only solution was to have an officer from each unit assigned to the port for several days, before each shipment, to track down the equipment of his particular unit.[166] Markings on cases and records of the various transactions were often inadequate, since shipments from the home stations to the ports were made by personnel inexperienced with overseas supply procedures, and Port Air Officers were unable to furnish detailed vessel

* A unit with 45 teletypewritten pages of shortages is cited in "Supply of AAF Organizational Equipment to the Army Air Forces," p. 35.

records, which were needed by the theater commanders and by Headquarters, ASC, to replace materiel lost at sea.

All of these factors pointed to the necessity for an automatic system of supply that would assemble and transport whatever was needed, where it was needed, with a minimum of wasted motion. But efficiency of supply was not the only reason for the introduction of a new system. Whenever equipment was moved so as to arrive ahead of its unit (whether at the Port of Embarkation or at some other station), that unit "was left with a job to perform and nothing to perform it with." [167] This was one of the prime considerations of the air force commanders who demanded permanent parent groups to accomplish "interim production." It was also a consideration for those responsible for technical training, who realized that the most effective preparation a group received did not come until its last few weeks in this country.

The movement of equipment by amateurs was particularly awkward in the case of Air Depot Groups, which in 1943 required approximately 125 freight cars for transportation by rail.[168] On numerous occasions the Air Service Command was criticized by the Inspector General for failure to have these organizations prepared for overseas shipment by specified commitment dates, but usually the criticism was that component units (especially Signal and Ordnance units whose equipment was "controlled") had not operated in the field with their own T/BA and OEL equipment.* On this basis, the Air Service Command felt, some units could never be considered trained. Air Depot Groups were not supposed to be mobile. They were semi-permanent establishments which were moved only in the event of unexpected strategical developments—as a

* The story of these tables is as follows. T/BA No. 1 for the AAF, dated 1 Oct 1941, was followed by a corrected table dated 1 March 1942. A new revised table was published 1 July 1942, to which changes were issued as of 25 Sept 1942 and 8 Jan 1943. Subsequently, instead of preparing new T/BA's, which listed items of equipment to be used by all types of squadrons in the AAF, it became the policy to prepare separate T/E's which listed equipment and supplies authorized for individual types of organizations often tabulated by kits, and after July 1943 T/O&E's which combined tabulations for both personnel and equipment. This latter table gradually became the standard publication for each separate unit, and was thereafter used as the basis for OEL's, which had formerly been based on T/O's, T/BA's and Tech Orders of the 00–30 series. (See ATSC Historical Monograph, "Supply of AAF Organizational Equipment to the Army Air Forces," pp. 29–30, and AR 310–60, 28 Aug 1943.)

result of changes in campaign plans rather than the progress of battle plans.[169] Control depots were the domestic installations which Air Depot Groups duplicated, and control depots in this country were not located at combat training stations. Furthermore, it was impossible to pack, box, load, unload, and set up the organizational equipment of these groups for field operations without robbing them of essential administrative, technical, and unit training.[170]

Another criticism of the supply system at this time was that the tables of organization and equipment were too rigid, and that the Air Service Command was insensitive to suggestions from the field for alterations. Field officers felt "neglected" and complained that "obvious faults in organization and equipment had become so apparent that they invited comment from disinterested components." [171] The explanation for this condition, according to Colonel Leo H. Dawson, was that the tables had been changed too frequently, rather than too seldom. While there was no question that they could be still further improved, "it is believed that the remedy is worse than the disease." [172] Constant changes created confusion even greater than that caused by obsolete tables.

In the spring of 1943 the Operational Training Unit (OTU) plan was announced whereby the shipment of permanent AAF equipment* was postponed until the organizations had reached their final phase training bases. The idea was that all OTU bases would be completely and permanently equipped so that tactical organizations would not need their own equipment before they arrived at final phase stations.[173] This system eliminated the necessity for last-minute shortage reports, and substantially reduced the shipment of equipment within the United States before delivery to the ports.[174] It gave supply agencies the maximum length of time between the activation of a unit and its final departure for assembling the necessary equipment, and provided each organization with new and serviceable equipment on the eve of overseas shipment. Furthermore, it precluded the cost of shipping great quantities of materiel from station to station in this country, saved the time required for crating and uncrating, and permitted closer, more systematic supervision over the distribution of equipment and the final packaging at the

* The ASC did not participate in an OTU supply plan for SOS (ASF) organizational equipment until 17 June 1944. See Daily Diary, Air Services Div (Supply & Services), AC/AS, MM&D, 17 June 1944.

port.[175] Above all, the OTU system of supply provided each trainee unit with a full set of organizational equipment at each point in its sojourn in this country.

As early as April 1943 the OTU system was being put into effect wherever possible. At an ASC training conference it was reported that "equipment for two Service Group Headquarters Squadrons was available at Dale Mabry; one set of Service Squadron equipment at Kelly Field; one set at Hunter Field; two at MacDill, Santa Maria, Pendleton, and Greenville, SGTC; and one complete set for an entire group at Will Rogers, Muroc Lake, Venice, and Fort Dix." [176] Each of these sets was put in under an assembly number; the base or sub-depot supply officer let the equipment out on memorandum receipt, and had it returned when the using units changed stations.[177] Meanwhile, OEL equipment for air depot training stations had been spotted as early as October 1942 at Albuquerque, Stinson, Warner Robins, and New Orleans. When these initial shipments were declared insufficient early in 1943, increased quantities of all types of supplies, including engine overhaul equipment,[178] were forwarded to the using stations. In May, General Frank reported that 75 percent of the OTU bases of the Air Service Command were fully equipped, and the balance could be equipped within 60 days.[179] His estimate was overoptimistic, however, for the stations were not fully equipped and the OTU system was not fully effective until after the approval of T/A 1-26, in September 1943.[180]

A feature of the centralized OTU supply plan was the assembling of organizational equipment at storage depots before it was issued to the using organizations.[181] OEL's prepared by the Air Service Command were assigned AAF "assembly numbers," each number representing a squadron of a given type. The sets of organizational equipment which were preassembled were prepared in two sections: the first was shipped directly from the assembly points to the Intransit Depots serving the Ports of Embarkation; the second, TAT (to accompany troops), was shipped to the final phase training stations. This policy enabled the Air Service Command to build up reserve stocks to meet demands for organizational equipment both in this country and overseas; it also gave the supply agencies more time for the orderly handling of shipments.

As an extension of this policy, some ports built up reserve sets of Air Corps equipment and kept them available for shipment to any theater at a moment's notice; the New York POE, for example, had twenty-eight sets, including Service Group and Air Depot Group equipment, ready in April 1943 for instantaneous shipment.[182] By August 1943 the Air Service Command was preassembling and occasionally preshipping AAF equipment for the following types of organizations:

Air Depot Groups
Airdrome Squadrons
Airways Detachments
Armored Divisions, Artillery Battalions (and other similar AGF organizations to which liaison type airplanes were assigned)
Bombardment Groups (Heavy, Medium, Light, and Dive Bomber)
Engine Overhaul Sections
Fighter Groups
Photo Squadrons
Service Groups
Troop Carrier Groups [183]

Preshipment of organizational equipment was expected to be of great assistance in the implementation of the Bradley Plan. The new system was thrown off, in this case, however, because while supply instructions were issued on the basis of squadron T/E's, Bradley Plan personnel were never organized into orthodox squadrons. Another complication was that many of the squadrons listed for activation in the United Kingdom were either entirely new, or else had T/E's that were then being revised.[184] To solve these problems it was determined that complete sets of specified squadron equipment should be released from storage at Kansas City, Missouri; Sioux City, Iowa; and Warner Robins, Georgia; and shipped to the theater forthwith in accordance with the best available estimates.[185] Special projects were then set up, and controlled items were expedited by priority extractions.

It was at this point that the system broke down, for the estimation of overseas requirements was weakest just when the system of shipments became most efficient. The result was a gross overshipment of critical and controlled items—with much damage to the domestic training program. Thus, when Bradley Plan requirements were revised in November 1943, it was found that a large excess had already been shipped and that this process was continuing.[186] For example, controlled items of Ordnance equipment for over 270 Mobile Repair Units were shipped to the United Kingdom alone in a period of only a few months.[187]

In still another manner, overzealous shipment to the theaters was responsible for the curtailment of the domestic training program. In some theaters service centers and air depots had been set up for two-shift or three-shift operations. In these cases the personnel of two or three Service or Repair Squadrons worked with the equipment of one such squadron, but often each of the squadrons was supplied with a full set of OEL equipment.[188] The elimination of unnecessary duplication should not have been too hard, but coordination with the other arms and services was slow, and for many months the shipment of organizational equipment based on the total number of squadrons, or supposed squadrons, produced a surplus of equipment in overseas Base Air Depots.[189]

★★★ 9 ★★★

# STANDARD PROCEDURES—NEW PROJECTS

## Special Projects: 1944–45

On 17 July 1944 the Materiel Command was united with the Air Service Command to form the provisional organization known as AAF Materiel and Services, Patterson Field, Ohio.[1] Lieutenant General William S. Knudsen, former President of the General Motors Corporation, was named Director; the Deputy Director was Major General Bennett E. Meyers. At the same time the Office of the Assistant Chief, Air Staff, Materiel, Maintenance, and Distribution, in Washington was redesignated Materiel and Services.[2] Effective midnight 31 August 1944, the combined commands were redesignated the Air Technical Service Command (ATSC).[3] The Office of the Chief of Engineering and Procurement was established at Headquarters, ATSC, to direct the operating functions of the former Materiel Command at Wright Field,* and the Office of the Chief of Supply and Maintenance to direct the operating functions of the former Air Service Command at Patterson Field, Ohio.[4] This reorganization of the central headquarters coincided with the appearance of several new and important projects connected with the training of service units.

During 1942 and 1943 emphasis had been placed on the creation of a unified national training structure which would coordinate the training programs of the several area commands. The process of standardization had been necessary, as regular procedures had to be worked out for all aspects of the program. Once centralized control had been established, however, emphasis shifted to the accomplishment of a number of urgently required special projects. To meet exceptional demands,

* Wright Field and Patterson Field are located between Dayton and Springfield, Ohio. They are separate fields, five miles apart. At the time of this reorganization the headquarters consisted of three parts: one at each field, and one half way between.

some units had to be readied with unprecedented speed; others prepared for unusual purposes. The purposes for which new projects were instituted included functional innovations, for which conventional Air Depot and Service Groups were inadequate, and racial projects, such as the establishment of separate Chinese and colored service organizations.

The development of special projects was thus a refinement of the later stages of the war. The very existence of such projects presupposed well-formed plans and known requirements, or else the introduction of novel technical inventions, like jet propulsion, necessitating immediate alterations in the organization and training of air service units. Previously the requirements of the theaters had not been so well defined. During 1942 and 1943 all groups had been trained as nearly as possible in accordance with a norm; there was no way of predicting where they would be used, or what type of combat units or aircraft they would service.[5] Not until the beginning of the VHB (B-29 and B-32) Project were service units trained for a particular type of aircraft with any assurance that that was the type they would be responsible for when they arrived overseas.[6] Even in 1944 it was not possible to predict the ultimate disposition of all service units; hence the necessity for special treatment in the case of priority undertakings. Another reason for special projects was that the existing AAF Troop Basis had made no provision for these units, and special adjustments had to be made to incorporate them.[7]

One of the first of the special projects of the Air (Technical) Service Command was known as Five Service Groups for CBI.[8] The occasion was the shortage of service units in India and China for the lines of the Air Transport Command.[9] To relieve this shortage it was planned that five old type Service Groups, including the 14th Service Group (Chinese), would be prepared in this country for immediate shipment to the CBI theater. The project was only partially successful, however, because the 14th Service Group (Chinese) took longer to train than had been anticipated, and an insufficient number of personnel was available for the manning of the other four groups at the time they were needed.[10]

A similar plan was the Smith Project, which called for the training of five Service Groups (Special) and two Air Depot Groups to accompany "commando" units in the same area; the units serviced were to consist partly of troop transports and partly of fighters and bombers. A new feature of this project was the creation of Air Cargo Resupply Squadrons,[11] which consisted of colored enlisted personnel,* and received their training at Lawson Field, Fort Benning, Georgia, under the auspices of the Troop Carrier Command. Training for this project was completed in the Warner Robins and the San Antonio areas,[12] but the units were not all used for the purposes for which they had been trained.[13] Some were substituted for other units intended for the CBI theater, however, and thus to a certain extent this project was a successor to the earlier one. At any rate the two projects together served in part to relieve the critical situation which had arisen in India and China.[14]

The Miracle Project at Fresno, California, consisted of the training of five Service Groups (Special) to accompany very long range (VLR) fighters, which were to be based on the Marianas and elsewhere in the Pacific to take part in the bombing of Japan.[15] These units were trained in the usual manner but in record time,[16] and shipped overseas to accomplish their purpose—except that two of them were later diverted from VLR (fighter) to VHB (bomber) units.[17]

Another special project was the training of a large number of individuals at San Bernardino for the Air Transport Command.[18] Approximately five thousand men were trained as first and second echelon mechanics (SSN 747). They were not organized as military units.[19] Most of the teachers were civilians, who had formerly taught third and fourth echelon subjects. There were also a few enlisted instructors. Trainees averaged less than one hour per day in classrooms; the rest of the time was spent servicing cargo planes, especially C-47's, C-46's, and C-87's. Indeed, the program was so practical that one of the chief duties of those in charge was to keep the Maintenance Division at San Bernardino from borrowing the men during work blitzes.[20] About sixty students per week were chosen to go to factory schools for training as aerial engineers.[21]

A later project at San Bernardino offered an even better illustration of the manner in which functional special projects were accomplished. This was the formation of a Service Group (Special) for jet-propelled P-80's and P-81's.[22] An early suggestion that a complement of approximately eight hundred specialists should be trained at one of the depots until a Service Group was required, and then transferred to tactical groups with a minimum of additional training, was rejected by the

* The first experimental Air Cargo Resupply Squadron sent to CBI was made up of white personnel.

Training Section, ATSC, as ill-considered.[23] It was pointed out that only a hundred specialists per group would be affected by the change to jet propulsion, and that the remaining seven hundred would provide ordinary supply, maintenance, and administration. Accordingly, plans were made for a normal three phase training period consisting of individual, unit, and combined training.[24] Existing Tables of Organization for standard Service Groups (Special) were altered only by the omission of propeller and engine mechanics of the conventional sort, and the substitution of jet-propelled engine mechanics. Jet tools were added to the standard Tables of Equipment in accordance with lists prepared by the Maintenance Division, ATSC.[25]

The group designated for the jet project was the 361st Service Group (Special). It was activated at San Bernardino, California, on 19 January 1945, and trained at that location until the middle of July, at which time it was transferred to Santa Maria, California.[26] The original cadre was trained at the Warner Robins School; San Bernardino acted as an OTU station and obtained T-A 1–26 equipment; specialists were sent away to technical schools in the usual manner.

Thus, standard procedures of organization and training were applied to the accomplishment of new special projects. This was particularly true of two large projects undertaken in 1944 and 1945 in cooperation with other branches: the Depot Unit (Army) Project, and the Army Aircraft Repair Ship Project. It was also true in the preparation of Chinese and colored service units.

## Functional Projects

### DEPOT UNIT (ARMY) PROJECT

This was a unique program involving close cooperation with the Army Ground Forces. It was carried out partly in this country and partly overseas. Altogether nine Depot Units (Army) were activated. Units 1, 2, 7, 8, and 9 were trained at Ogden ATSC, Utah; the rest were trained in England.[27] These units were needed because of the increasing use of liaison aircraft by the Field Artillery Corps of the Army Ground Forces, especially in the Italian theater of operations.[28] In the past the Army Ground Forces had maintained its own liaison aircraft, turning them over to ASC (ATSC) installations only for top echelon repair.[29] Since these installations accorded priority to combat airplanes, however, the

Army Ground Forces was dissatisfied with the service it was receiving, and proposed that a new maintenance unit for liaison aircraft be created under AGF supervision. If this plan had been approved, the Army Air Forces would have been connected with the project only in the matter of supply.[30]

The Army Air Forces advanced a counterproposal that the activation and training of these units should be performed under AAF supervision, and the unit should be a conventional, though reduced and modified, Service Group for the maintenance and supply of Ground Force liaison aircraft. It would serve under the operational control of the Ground Forces, but under the technical control of the Air Service Command. One of these units would be assigned per army, or in exceptional cases one per separate corps. This was the proposal which was finally accepted by G-4 (Supply) of the War Department.[31]

Coordination between the Army Air Forces and the Army Ground Forces was accomplished by a series of conferences at Patterson Field, Ohio,[32] at which basic decisions were made on manning, equipping, and training. The plan that was formulated provided for a third echelon maintenance and supply organization to service all types of liaison aircraft, especially L-4's and L-5's. There were to be 4 officers and 54 enlisted men in each unit,[33] and the equipment was to be such that the unit could perform its mission in the field and still be mobile enough to accompany moving elements of the Ground Forces. As a result of these preparations, five of the new units were activated at Ogden ATSC in 1944 and 1945.[34]

The training program for Depot Units (Army) was entirely conventional.* Individual military and technical training was followed by unit and combined training, just as it was in the training of Air Depot and Service Groups.[35] On O-Day, the date the units reached 90 percent strength, basic military training was begun. Shortly afterwards the units participated in a four-day bivouac at Camp Williams, Utah. While on bivouac, they performed their own administration, established interior guard, practiced camouflage discipline, and participated in field problems involving map reading, convoy driving, and security. Special emphasis was placed on this phase, for the units were later to be transferred for combined training to the Army Ground Forces, and it was hoped that they would give a good account of themselves. After the bivouac,

* For comparison with conventional units, see Fig. VII, p. 159.

the units returned to Hill Field, Utah, by motor convoy to begin basic technical training.[36] In technical training the personnel were assigned to post schools and supervised by civilian instructors. Following four weeks of supervised training, they were allowed to operate on their own. During this period they were trained on Class 26 materiel.*

Practical experience during unit operations exposed a number of errors in organization.[37] It was observed, for example, that anything less than a thirty-day level of supplies was insufficient, since few of the continental air depots kept regular stock levels of supplies for liaison aircraft. It was also noted that there was an insufficient number of dope and fabric workers and aircraft and engine mechanics; there were only five 548's and seven 747's, although both of these classifications were essential. To offset this condition a program of cross-training was initiated to teach personnel to perform duties other than those indicated by their primary classifications.

Upon completion of unit training at Ogden, the units were assigned to Army Ground Force stations or to Army Air Bases throughout the country to maintain liaison aircraft assigned to the various elements of the Field Artillery and/or to Liaison Squadrons. Difficulties discovered during combined training were discussed at on-the-spot conferences attended by the Ogden ATSC Liaison Officer, the Air Artillery officers of the Second and Fourth Armies, and representatives of the Assistant Chief, Air Staff, Training.[38] Last-minute changes in personnel and equipment were thereupon incorporated into the appropriate War Department directives.[39]

The 1st Depot Unit (Army) was sent for final phase training to Camp Hood, Texas; the 2nd to the 415th Field Artillery Group at Fort Jackson, South Carolina, thence to the XXXII Corps Field Artillery, at Fort Bragg, North Carolina, and thence to Lafayette Army Air Base, Lafayette, Louisiana; the 7th to Fort Bragg, North Carolina; and the 8th to Brownwood Army Air Base, Brownwood, Texas.[40] In each of these areas the units serviced from forty to seventy liaison aircraft, which were operating from widely dispersed flight strips as participating elements in Army Ground Force maneuvers.[41]

* Class 26 materiel, equipment used for instructional purposes, was obtained from the Supply Divisions of the area commands, or from the Supply Division, ATSC.

### ARMY AIRCRAFT REPAIR SHIP PROJECT

The Army Aircraft Repair Ship Project was the Army Air Forces' answer to the demands of the Pacific theater of war for mobile, sea-going Repair and Maintenance Units.[42] As the island-hopping campaigns of the Pacific accelerated, it was seen that conventional, land-based air depots and service centers were altogether too cumbersome. By the time an old-style depot could be packed up and moved from one island to another, the "front" would have advanced so far that the new installation would be useless even before it was fairly established. Accordingly, with the permission of the U.S. Navy, six 10,000 ton Liberty Ships and eighteen smaller vessels were converted into floating facilities known respectively as Army Aircraft Repair Ships and Army Aircraft Repair Ship Auxiliaries.[43]

The larger Repair Ships were designed to perform fourth echelon supply and maintenance, exclusive of engine overhaul, for several groups, depending on the type of aircraft serviced. The smaller Auxiliaries performed third echelon work for approximately the number of airplanes assigned to one group. The units assigned to these ships were known as Army Aircraft Repair Units (Floating) and Army Aircraft Maintenance Units (Floating). Most of the work was to be done in the shops and on the decks of the ships, but smaller workboats, jeeps, ducks, and helicopters were also provided to enable service personnel to go into otherwise inaccessible places with enough equipment for spot repairs.[44]

As far back as 1942 the Army Air Forces had anticipated the need for mobile repair units in the Pacific, and the Director of Base Services, Colonel L. P. Whitten, had requested fifteen 250′ boats to be used as aircraft tender and supply boats.[45] Early in 1943 the Commanding General, Air Service Command, had requested that consideration be given to furnishing several large "floating air depots" with equipment comparable to that of Air Depot Groups.[46] But these suggestions had never been carried out. The real impetus for the project came from a survey of the South Pacific Area in the fall of 1943 by an inspection party which included Lieutenant General Brehon H. Somervell, Commanding General, Services of Supply; Major General Oliver P. Echols, AC/AS, MM&D; Major General Charles P. Gross and Brigadier General J. M. Franklin of the Transportation Corps; and Colonel Paul E. Ruestow,

Chief, Logistics Planning Branch, AC/AS, MM&D.[47] These officers were everywhere struck by the inadequacy of the existing arrangements and the apparent necessity for some mobile means of aircraft repair. Furthermore, they were favorably impressed by the practicality of a naval repair ship for surface craft, which they visited at Perth, Australia, and felt there was a definite need for a similar floating maintenance unit for the Army Air Forces.[48]

On their return they asked the Air Service Command for recommendations upon which to base a request for a floating maintenance unit of the type envisaged.[49] The Air Service Command suggested that several large ships and eighteen or so smaller ships would be more useful than one very large ship.[50] General Echols approved the proposal for Headquarters, AAF,[51] and on that basis the Air Service Command immediately started to accumulate the necessary equipment without waiting for specific authority, thereby saving many months.[52]

There was considerable delay in the authorization of the ships, however, for when the allocation of these vessels was brought before the Joint Chiefs of Staff, Admiral E. J. King, Commander in Chief, U.S. Fleet, objected to the proposal on the ground that the Navy's experience with floating facilities of this type had not been fully taken into account.[53] Furthermore, in subsequent conferences it developed that the Navy did not like the idea of the Army Air Forces' going to sea. General Whitten, for the Army, pointed out that no objections had been raised to the Navy's programs for shore-based establishments and Seabee Battalions, "although to some of us the quantities appeared excessive," and finally after a number of details had been worked out the Navy agreed to cooperate.[54] At the direction of Admiral William D. Leahy, the President's Personal Chief of Staff, six Liberty Ship (EC2-S-CL) hulls were turned over to the Transportation Corps at an early stage of completion for conversion into Repair Ships,[55] and at the direction of Major General Lucius D. Clay, Director of Materiel, ASF, eighteen 178′ craft were diverted for use as Auxiliaries from a number then being mass-produced at the Higgins Shipyards, New Orleans, Louisiana.[56]

Even before Admiral Leahy's "green light," a committee had been appointed by Major General Clements McMullen, Chief of the Maintenance Division, ASC, to expedite the conversion of the ships, acquire equipment and supplies, and procure and train the personnel. This key group was largely responsible for the swift development of the project.

At Headquarters, AAF, the focal point was the Office of the Marine Section, Air Services Division, where Major David D. Lent handled all matters relating to the conversion of the vessels and the subsequent development of other phases of the project. To assist him in keeping the Air Service Command informed on all details a Liaison Officer, Lieutenant Colonel W. R. Weber was assigned to Headquarters, AAF. At Mobile and San Antonio ASC's Special Project Officers were appointed to supervise the organization and training of the tactical units which manned the ships.

Conversion of Repair Ships started in April 1944 in the yards of the Waterman Steamship Company and the Alabama Drydock Company in Mobile, Alabama.[57] The original layout of the vessels had been prepared by the Air Service Command on the basis of Navy drawings, but the drawings had to be changed because the ships' crews were to be civilians of the Merchant Marine, and the latter organization, unlike the Navy, forbade the quartering of crew members below the water line. This requirement had not been foreseen by the ASC Project Engineer, but the change was quickly made, and the conversion proceeded. The necessity for priorities, fortunately, was avoided. Although no actual priority was ever given to the Army Aircraft Repair Ship Project, the Air Services Division, MM&D, Headquarters, AAF, wrote to the War Production Board and the War Manpower Commission in Washington and to their branch offices in the Southeastern District to the effect that the project was to have a "must" rating, at the direction of the Joint Chiefs of Staff, and this informal request was scrupulously honored.

Auxiliaries required more conversion than Repair Ships because extensive changes had to be made in the physical outline of the ships' structures.[58] The craft procured had not been intended for the use to which they were now being put; Landing Craft Tanks (LCT's) had been the Air Service Command's first choice for this purpose, as they had flat bottoms and could be landed in very shallow water. It had been impossible to pry any LCT's loose from the Navy, however, so that idea had been abandoned. Other delays were experienced in obtaining equipment, especially generators, but these were finally diverted from other uses and the ships were soon made operational. Some of the delays in obtaining equipment were caused by the lack of a clear line of demarcation between the responsibilities of the Army Air Forces and those of the Transportation Corps of the Army Service Forces.

The aim in equipping both Repair Ships and Auxiliaries was to provide "front line" workshops, where spare parts could be made, rather than to attempt to supply ready-made all the thousands of parts which might eventually be needed.[59] The larger ships, in particular, were fitted with many shops for many purposes; the most important being the machine shop, the oxygen plant, and the rubber repair shop. The supplying of the ships was speeded by the fact that the Air Service Command had anticipated the need for great quantities of supplies, and had already made them available.[60] On the other hand, there had been no advance knowledge of the types of aircraft to be serviced, other than B-29's, so there was inevitably a good deal of guessing. Because of the relative value of the ships, and because they would probably be used within range of enemy bombers, they were well prepared for defense.[61] In addition to sixteen large dual-purpose guns and any number of small arms, they were equipped for defense against gas and incendiary attacks, and all personnel were issued naval gas masks.

The manning of both Repair Ships and Auxiliaries presented unusual complications, for there were to be Navy personnel and civilians aboard, as well as AAF technicians.[62] Moreover, it was necessary to cut all three categories to a minimum, so that there would be some room on the crowded ships for work. The crews were to be civilians, paid 100 percent premiums for their risks. The gun crews were to consist of both Navy and Army personnel, with the understanding that the Army might assume this responsibility completely, once its personnel were sufficiently well indoctrinated. The Navy had insisted from the beginning, however, that the protection of these ships was strictly a Navy function.

So far as AAF personnel were concerned, the ships' complements were to be economical in the extreme. As originally contemplated the program called for six Repair Ships and thirty-six Auxiliaries. The former required 22 AAF officers and 340 enlisted men each; the latter 3 officers and 48 enlisted men each. This was a total of 240 officers and 3,768 enlisted men for the whole project, to be obtained by substituting the floating units for four Engine Overhaul Squadrons in the AAF Troop Basis for 1944.[63] But, in spite of the relatively small numbers, it was not possible to obtain a sufficient quantity of technicians in the several MOS's, and the number of Maintenance Units (for the Auxiliaries) had to be reduced to eighteen, bringing the total for all vessels, large and small, to 186 officers and 2,904 enlisted men. This was approximately the number sent to the Pacific.[64]

The organization of the units was different in that there was only one element of command for each unit in the project.[65] By this device the number of administrative personnel, including company officers and first sergeants, was greatly reduced. Each unit was divided into Administrative, Tactical, Service, and Technical specialties, a system of organization not unlike that of the new Service Group (Special). Unfortunately, the original organization of the units had been hastily and inaccurately calculated, and there were some cases in which the T/E's provided equipment for which there were no workers in the T/O's, and vice versa. For example, maintenance equipment for the calibration and repair of bombsights was included, although there were no bombsight specialists (SSN 574) in any of the units.[66]

In order to train the men for floating maintenance and repair, a five (later, six) months' training program was set up embracing individual technical and specialized training, marine training, mock-up training, training at special schools, and shipboard training.[67] Most of the courses were conducted at or near Mobile, Alabama, where good weather, plenty of navigable water, and well-equipped shops and docks provided excellent facilities for Repair Ship training. Brookley Field at the Mobile Air Depot was designated an OTU station, and authorized to draw T-A 1–26 equipment for this purpose.[68]

At the beginning of the program all men received technical training at the Brookley Field post schools. This was a poor idea, as many of the men had had the same courses before—too frequently the case in military training. Later an arrangement was made whereby qualified individuals were allowed to omit school training and substitute for regular employees in the shops of the Maintenance Division. This procedure enabled the men to gain practical experience, and their experience was varied whenever possible by moving them from one part of the line to another. One of the courses given at Brookley Field at this time was for physical fitness experts and Special Service (recreation) Officers. A good deal of equipment was provided, since men living in cramped quarters, away from land for long periods of time, would need extra facilities for recreation.

Marine training was conducted at the Grand Hotel, south of the city of Mobile on Mobile Bay (Gulf of Mexico). Here the men were given instruction in swimming, life-saving, elementary seamanship, ship's customs, and the handling of small boats and cargo. The school at this location was staffed by personnel from the Navy, the Army Air Forces, the Merchant Marine, and the Transportation Corps of the Army Service Forces.

After marine training the men were sent to Bates Field, Alabama, near Mobile, for mock-up training in complete, made-to-scale replicas of the various shops that would be found on board ship. The purpose of this period was to give advanced technical training, and to familiarize the men with the crowded conditions under which they would be working. The only trouble was the shortage of repairable parts and Class 26 equipment.

For specialized training the men were sent to a number of schools in various parts of the country. For example, some were sent to the Naval Armed Guard School at Norfolk, Virginia, to learn the operation and care of Navy guns. Others were sent to Freeman Field, Indiana, for instruction in the piloting of helicopters, and still others to Bridgeport, Connecticut, for the repair of helicopters. A few were sent to Kelly Field, Texas, and Robins Field, Georgia, for radio and radar work, and several others to Chattanooga, Tennessee, to learn to operate oxygen manufacturing plants.

In theory the men were to receive a month of shipboard training on Mobile Bay. However, the units were believed to be so badly needed in the Pacific that shipboard training was given on the way over. The first of the Repair Ships to go left in September 1944. The last unit in the project was shipped on 6 February 1945.[69] A total of six Aircraft Repair Units (Floating) and eighteen Aircraft Maintenance Units (Floating) was shipped as planned, with 186 officers and 2,874 enlisted men.[70]

## Racial Projects

### CHINESE SERVICE UNITS

A special project of a different sort was the training of the 14th Service Group (Chinese) at Venice Army Air Base, Venice, Florida.[71] This organization was established with the idea that it would eventually be used in China; it was a conventional, old-type Service Group no different from any other, except that many of its officers and all of its enlisted men, but one, were Chinese. That one was a staff sergeant who served as assistant engineering officer. The group consisted of a Headquarters Squadron, one Service Squadron (the 555th, which had formed the original cadre for the group), two Quartermaster Truck Companies, one Quartermaster Company (Service Group), and two Ordnance Supply and Maintenance Companies (Avn).

The 14th Service Group (Chinese) was trained in the Warner Robins area, in Georgia, under the supervision of Colonel Louis A. Merrillat, Special Training Coordinator for Headquarters, ASC, and the group's commanding officer, Colonel Eric H. Kaeppel, former Chief of the Personnel and Training Division, WRASC. The Executive Officer was Major Woodrow Wilson Lee (Chinese). In September 1944 there were fifty-four officers in the units that made up this group, and of these, twenty-one were Chinese, two were Korean, and the rest were American whites. Most of the Chinese officers had been born in China, but educated in this country; all of them spoke English.

Approximately 85 percent of the enlisted personnel had also been born in China; for the most part they, too, had learned to speak some English. They spoke several dialects of Chinese, however, depending upon their local origin, and this caused the Air Service Command no end of trouble. It was actually necessary to teach these men to speak "Chinese," that is, one dialect of Chinese, or Mandarin, before the group could begin training. Language classes had to be set up for both officers and enlisted men, and a manual on the subject, "The Fundamentals of Chinese Mandarin," had to be published to make it possible for the men to communicate with each other in Chinese as well as in English.[72] Furthermore, to avoid possible misunderstandings, the American Articles of War were published in Chinese for the benefit of all concerned.[73]

In February 1944 technical schools were set up in the group area, and all personnel were sent to these schools to complete their technical training. Instructors from the 27th Parent Service Group, recently redesignated the 4500th AAF Base Unit, supervised all instruction, and classes were held in machine shop work, welding, propellers, instrument repair, woodwork, parachutes, and technical supply and administration. Airplane mechanics were given on-the-job training at various locations in the Florida area; Class 26 materiel was procured from Warner Robins ASC. Specialists of various sorts were sent on detached service to schools throughout the country. So far as the general organization of training was concerned, this group was not exceptional in any way. There were, however, certain differences.

One of the problems was the matter of obtaining American citizenship. At the beginning of the training period only 60 percent of the enlisted men were U.S. citizens. Moreover, there were several cases of illegal entry into the United States. Citizenship classes were set up, and soon afterwards weekly pilgrimages were conducted to the Federal Court at Tampa, Florida. But legal hindrances were encountered because some of the applicants could not remember their ports of entry, or were suspected of giving any port regardless of the truth. This requirement delayed some cases for several weeks, until immigration authorities had satisfied themselves that they had obtained the exact port of entry. Another difficulty was that while Chinese ordinarily gave their family names first, some of the men had reversed the process in order to Americanize their names.

Administrative snags were likewise encountered in obtaining allotments for dependents. Some of the soldiers' families lived in Japanese-held territories; others came from homes which might or might not have been in Japanese hands. Several claimed no dependents at all when they entered the service, but later stated that they had wives, children, parents, grandparents, and other dependents. These cases all had to be investigated, and a number were kept pending for a long time because of delays in locating families in a country where large migrations had taken place. Investigations of dependents in China were made through the Red Cross and the Chinese Embassy in Washington. It was found that there were several instances in which a soldier's pay had been deducted for an allotment for almost a year, and yet his family had received no payment during the entire period. In these cases, the money was refunded.

The physical characteristics of the Chinese people introduced a novel technicality, in that many of the men were underweight by American standards. Consequently, they had been declared unfit for overseas duty. But regardless of their slight frames and short stature, it was demonstrated on many occasions that they had amazing endurance under strain and were fit for all types of general duty. The regulations were relaxed in some cases, therefore, and only those who were definitely unqualified were weeded out and transferred to other units.

At the insistence of the Special Training Coordinator from Headquarters, ASC, the policy was instituted of serving garrison rations consisting of Chinese food. Under this system rice took the place of bread and potatoes, and the food was cooked in Chinese fashion. "Chow-Gow-Woo," or tripe,* was often served, and "Bock-Choy," a leafy vegetable,* was obtained from various farms in Florida. However, the majority of men in the group preferred American to Chinese dishes. Some had acquired a taste for American food in civilian life; others had formerly been assigned to American groups and had become accustomed to ordinary GI chow. Thus, there was a good deal of complaint that Chinese food, prepared in bulk, was not to their liking. Whether or not this type of cooking was good training for the future remained to be seen.

The religious affiliations of the personnel of this group, including both officers and enlisted men, were listed as follows:

| *Religion* | *Percent* |
|---|---|
| Protestant | 18 |
| Catholic | 2 |
| Jewish | 1 |
| No denomination | 79 |

Among those who listed themselves as having no denomination, it was estimated by the group historian that approximately 28 percent of the total were Confucianists (not a religion) and 5 percent were Buddhists. Usually about 10 percent of the men attended Sunday services, which were conducted in the ordinary manner. No services or readings were held for Confucianists or Buddhists.

Unusual situations of a different sort arose in the day-to-day contacts of officers and men. American officers accustomed to American soldiers were as likely as not to receive the surprise of their lives when Chinese GI's responded to orders by attempting to make bargains. If an officer assumed an overbearing attitude and threatened to place a soldier in the guardhouse, the soldier immediately responded with an attitude of stoical indifference, and seemed to prefer the guardhouse to remaining under that officer's command. Exasperation came easily, but if the officer forgot his position with these men, he lost face. Once he lost face, he could get nowhere.

It was the opinion of the commanding officer of the group, Colonel Eric H. Kaeppel, according to the group historian, Staff Sergeant John

* The translations were supplied by the group historian.

S. Stuart, that the chief reasons for low morale and laxity in military discipline among the personnel of the group were:

(a) Inexperienced officer personnel;
(b) Insufficient basic training;
(c) Lack of the right type of orientation.

Orientation, in particular, the group historian felt, had been a weak point in the over-all training program. It was not sufficient, he said, to present such motion pictures as "The Battle of Britain" to orient these men; yet that was the type of orientation they received. ("The Battle of China" was not available at the time this group was trained.) [74] It was more important that lectures be given in simple English (or Chinese) by someone in authority, and that the lectures be followed by a question period when answers might be given to such questions as: "What does it mean to be an American?" "Why are we in the war?" and above all, "Why has a separate Chinese Service Group been formed?" The last question was the hardest to explain, as most of the men felt that their opportunities for promotion and advancement had been better in American units than they were in the 14th. Also, they felt that they would be working more efficiently and would have more incentive to put forth their best efforts in American units than in an all-Chinese unit. As a result of this attitude, repeated requests for transfer were directed to the Air Inspector at Warner Robins—from both officers and enlisted men.[75]

### COLORED SERVICE UNITS

Long before the beginning of World War II the policy of the War Department for the treatment of colored troops had been premised on the belief that these troops were a separate problem. Mobilization Regulations issued in July 1939 [76] summarized the Army's long-standing policy on colored personnel as follows:

> . . . The War Department will indicate in mobilization plans A PERCENTAGE (OF THE TOTAL MOBILIZED STRENGTH) AT WHICH NEGRO MANPOWER WILL BE MOBILIZED AND MAINTAINED FOR EACH PERSONNEL PROCUREMENT PERIOD. The percentage will be approximately equal to the ratio between the negro manpower of military age and the total manpower of military age. . . .[77]
>
> (3) Unless conditions require modification in the interests of national defense, the ratio of negroes mobilized in the arms, as compared to those mobilized in the services, will be the same as for white troops. . . .[78]
>
> b. *Negro officer candidates.*—The maximum number of negro officer can-

didates included in requirements for that class of personnel will not exceed the number required to provide officers for organizations authorized to have negro officers, including loss replacements therefor, due account being taken of negro officers already available on initiation of mobilization. . . .[79]

Regulations concerning the mobilization of colored troops were changed from time to time, but the policies stated above remained in effect for all levels of the War Department. Emphasis continued to be placed on the mobilization of colored personnel in each branch of the service, combatant as well as non-combatant,[80] and on the requirement that colored soldiers be given an opportunity to qualify for reserve commissions through assignment to Officer Candidate Schools.[81] In subsequent directives particular emphasis was placed on the establishment of separate colored tactical units (as distinguished from "overhead installations," or headquarters) and the assignment of colored officer personnel thereto. Special provisions were made to facilitate this procedure. For example, it was stated in a War Department letter to all major commands that:

Commanders are authorized to waive age in grade restrictions to permit the assignment of qualified over-age negro officers to troop units.

Negro medical officers and chaplains may be assigned to a unit without regard to the fact that negro officers of the basic arm or service have not yet been assigned to that unit.[82]

Within the Air (Technical) Service Command, however, colored personnel were utilized during the war only in two Service Groups and in a limited number of Aviation Squadrons and ASWAAF units, including Quartermaster Truck Companies, Ordnance, Signal, and Chemical Warfare Companies, and Guard and MP Companies. Prior to August 1943 they were also utilized in Air Base Security Battalions, which were usually trained by the numbered air forces, but attached to Air Depot and Service Groups in the theaters of operations.[83] After August 1943, Air Base Security Battalions were inactivated as a result of the McNarney Directive. One of the reasons for the inactivation of these units was General McNarneys' conviction that they constituted the best source of high calibre colored enlisted personnel for the manning of tactical service units.[84] Following the inactivation of Air Base Security Battalions, 93 officers, 2 warrant officers, and 5,275 enlisted men (colored) and 38 officers, 4 warrant officers, and 898 enlisted men (white) were transferred to tactical service units of the Air Service Command.[85]

The two colored Service Groups activated during World War II were the 96th Service Group, which was trained at Oscoda and Mount Clements, Michigan, and the 387th Service Group, which was trained at Daniel Field, Augusta, Georgia. A great deal of difficulty was experienced by the Air Service Command in the training of both these groups, and the 96th in particular was considered "sub-standard." [86] On 4 August 1943, Brigadier General Elmer E. Adler, accompanied by staff representatives of the Training and Operations Section of the Air Service Command and members of the Personnel and Training Division of the Fairfield ASC, inspected this group at Oscoda, Michigan, and noted certain discrepancies in the administration of the unit which were attributed to the policies of the First Air Force. These discrepancies were described by General Adler, as follows:

This visit revealed that there was not sufficient Base personnel, thereby making it necessary for certain units of the 96th Service Group to perform Base functions, denying these units an opportunity to take full advantage of a very short training period. . . .

The Truck Companies at Oscoda were being required to perform Base functions two weeks out of every month. Such an arrangement is a definite handicap in the training of the personnel of these Truck Companies. . . .

In view of the above condition, it is requested that the First Air Force furnish personnel necessary to perform normal Base functions; such as, operating base administrative vehicles or performing base administrative duties. It is also requested that the Base Commander be directed to comply with the agreements previously arranged by this Command and Representatives of the First Air Force, and turn over to the Associated Arms and Services those functions which a Service Group is organized to handle.[87]

The agreement between the Air Service Command and the First Air Force, to which General Adler referred, provided that the base commanders of air force stations were to be responsible for all base functions not normally performed by trainee tactical Service Groups.[88] Since this agreement had not been complied with in the case of this organization, the 96th Service Group, which was supposed to be servicing the 322nd Fighter Group at Selfridge Field, Michigan (the group it was to accompany overseas),[89] had been compelled to perform base tasks not ordinarily included in the training of OTU units. Thus, the deficiencies found in the training of this organization were not entirely the fault of the personnel of the group.

There were, however, weaknesses which were attributed directly to the shortcomings of certain of the group's officers. In particular, General Adler stated, "there is lacking at the present time the proper senior colored officers who can command and staff the organization." [90] Accordingly, he recommended that a number of the colored officers then in charge be relieved, and that they be replaced by other, better qualified colored officers, if available, or by white officers "capable of commanding negro troops which will include negro Lieutenants and Captains." [91] General Adler's request was approved by the Deputy Chief of Air Staff, Brigadier General Edwin S. Perrin, who wrote: "The Commanding General, Air Service Command, is authorized to assign to the 96th Service Group (Colored), Oscoda, Michigan, qualified white officers in field grade. . . . It is imperative that this matter receive no publicity and that the negro field officers relieved from assignment be frankly and tactfully advised that the action taken by this Headquarters is necessary for the best interests of the unit itself as well as of the government." [92]

Since there were fewer than twenty colored officers in the entire Air Service Command, adequate colored replacements for the field grade officers who had been relieved could not be obtained, and the supervision of the 96th Service Group was turned over to a staff of white officers furnished by Headquarters, ASC.[93] Certain colored junior officers were also replaced at this time by white personnel. Unfortunately, the junior officers of the group were all recent OCS graduates, who had had very little Army experience and were not particularly skillful in the administration of their units. Consequently, they had allowed conditions to develop which were bad for the morale of their men,[94] and had condoned favoritism in the distribution of passes and furloughs. The reason for this, according to the group Chaplain, was that the sergeants were more familiar with the paper work of the unit than the newly commissioned officers, and if the officers had tried to interfere with what the sergeants were doing, the latter had threatened "to ball up the works." [95] If left on their own, the officers were unable to see that the paper work was properly accomplished. This condition had served to hamstring the administration of the group, and affect the morale of the men, which was already low because of the severe restrictions placed on all colored soldiers in the vicinity, following the recent race riots in Detroit.[96]

Another difficulty was that constant restrictions had caused a large number of AWOL's, and therefore an unusually high court martial rate.[97] Conducting court martials had increased the administrative responsibilities of the already overburdened officers.[98] The number of prisoners was considered excessive, but little could be done to cut it down, except to discourage new offenders. To accomplish this end, the recently assigned commanding officer of the Oscoda Army Air Field made certain changes in the guard house "by removing the mattresses and giving each prisoner sufficient work to make his sentence a penalty instead of a loafing period." [99]

The circumstances affecting the morale of the group, however, had not all been connected with military discipline. Visiting inspectors found that the attitude of the local civilians had played a definite part in creating the tense situation which had developed. The residents of Oscoda had objected to the presence of such a large group of colored soldiers in their community, and had publicly requested that they be removed. Furthermore, they had wanted to know why the Army had not been able to send a proper proportion of white soldiers, instead of such a great mass of colored men for whom there were only inadequate facilities. They had protested that such a large group of colored soldiers was too much for any average community to be expected to accommodate. This situation was described by Major Earl H. Devanney of the Military Personnel Section, ASC, as follows:

We next interviewed Mr. Lloyd Soucie who is Secretary and Treasurer of the local U.S.O. Committee and Commander of the local American Legion Post. Mr. Soucie said in part that "When the colored troops arrived in Oscoda the County Board of Supervisors of Iosco County passed a resolution and sent it to their Congressman asking that the colored troops be removed. This was given a great deal of publicity in the Pittsburgh Courier and other colored newspapers throughout the country. These papers are read by the troops and created quite a tense situation." He said, however, that the townspeople now have for the most part accustomed themselves to the presence of negro troops.[100]

The 96th Service Group was thus a cause of considerable administrative difficulty to the Air Service Command during most of its stay in this country. As a result of vigorous action and close supervision on the part of the highest authorities, however, the group was finally prepared for its POM inspection, and early in January 1944 it departed for the Port of Embarkation at Hampden Roads, Virginia.[101] It was not until July of the same year, approximately seven months later, that the next colored Service Group, the 387th, was officially activated.[102] And it was not until October 1944 that the personnel of this organization were assigned.[103]

The 387th Service Group at Daniel Field, Georgia, in contrast to the 96th, was officered from the beginning by white personnel.[104] One of the chief handicaps of this organization was that the majority of its enlisted men were lacking in basic military training, and that many of them were experiencing their first opportunity for assignment to a definite tactical organization.[105] Prior to this time they had been assigned as casuals at various stations, and had been required to perform a number of laboring tasks, especially in officers' clubs and BOQ's, without benefit of any sort of military training "except that they had been taught to salute." Inasmuch as the average length of service as of 15 December 1944 was 20.56 months, "this fact in itself has created a definite problem of morale resulting in skepticism still existing and being noticeable primarily in their attitude toward basic and technical training." [106] It was hoped that the skepticism could be overcome once the group had been transferred to a location where the personnel could work with a Bombardment or a Fighter Group, and the men could see the importance of their jobs in relation to a tactical organization.[107]

In general, the 387th was considerably less of a problem than the 96th had been. Several bivouacs were conducted, with emphasis on the procedures of convoy driving and security for all personnel, and numerous individuals were shipped to special schools for courses of technical training.[108] Early in December primary MOS's were certified, and later in the month an inspection was held to determine the fitness of the group for transfer to the jurisdiction of the First Air Force. This inspection was passed without difficulty,[109] for the group was not unaccustomed to inspections. In fact, by then the organization had been inspected by so many people that the group historian remarked that the men had come to feel that they were probably the most inspected group in the Army Air Forces.[110]

In view of the many comments of visiting officers that this Service Group was composed of better than average personnel, a survey of AGCT (intelligence) scores was conducted to arrive at a definite conclusion.[111] The average AGCT score of the 481 enlisted men, whose average age was 25.8 years, was found to be 88. This figure had little value, however, since no information was available on other, similar organizations with which to make comparisons. In order to make a somewhat more comprehensive statement, the scores of the personnel of the 387th Service Group were compared with those of all colored men processed at army reception centers during the period 1 January 1943 through 30 June 1943. The results of this survey are given in Table 12.

TABLE 12

AGCT GRADES OF COLORED PERSONNEL *

| *WD Pamphlet 20–6* | I | II | III | IV | V | *Total* |
|---|---|---|---|---|---|---|
| Number | 419 | 5,991 | 23,402 | 83,104 | 61,023 | 173,939 |
| Percentage | 0.2 | 3.4 | 13.5 | 47.7 | 35.1 | 100 |
| *387th Serv GP* | | | | | | |
| Number | 5 | 94 | 129 | 207 | 46 | 481 |
| Percentage | 1.4 | 19.54 | 26.6 | 43.4 | 9.52 | 100 |

* The Army General Classification Test (AGCT) was given to all personnel upon entry into the army as a means of measuring intelligence. The highest possible score was 163. Officer candidates had to score at least 110. Under 60 was generally considered "illiterate." Results were broken down into groups, as follows:

| | |
|---|---|
| Group I | 130 & above |
| Group II | 110 – 129 |
| Group III | 90 – 109 |
| Group IV | 60 – 89 |
| Group V | 59 & below |

The training of non-technical colored units, such as Aviation Squadrons, was sometimes accompanied by even greater difficulties than the training of Service Groups. The mission of Aviation Squadrons, according to AAF Training Standards, was simply the performance of the normal labor tasks connected with the upkeep and operation of army air bases.[112] Unit after unit, however, reported that the men had noted that these squadrons were reserved for colored troops. In some cases, descriptions of their attitudes were expressed in guarded language. For example, the historian of the 473rd Aviation Squadron at Kelly Field, Texas, expressed his thoughts as follows: "Due to the fact that the Squadron's mission is not of a very high technical nature, problems, with the exception of personnel matters, were kept to a minimum." [113] Some of the enlisted men frankly resented the use of a "highfallutin" name; it seemed to them that the term Aviation Squadron was a window-dressing to fool the public. "If they're going to put us in labor battalions," said one colored soldier, "why don't they at least call them labor battalions?" [114] Others disliked the language of the AAF Training Standard, which stated that the personnel of these units would be trained to be "competent to be of assistance about the hangar line." [115]

As a result of poor morale, Aviation Squadrons and other colored units often reported cases of "AWOL," "drunk and disorderly," "failure to obey orders," and "the use of insulting language."[116] Moreover, there was a high incidence rate of venereal disease (119 per thousand per month for colored troops, as opposed to 29.6 per thousand per month for white troops, according to General Beau).[117] Breaches of discipline were particularly serious in the South. The Commanding Officer of the 456th Aviation Squadron at Daniel Field, Georgia (a unit subsequently transferred to San Bernardino, California) described his experiences in the administration of his unit as follows:

> As is to be expected, the major organizational problem faced by the 456th was the morale situation under conditions of racial discrimination existing in the state of Georgia. Justifiable resentment at "Jim Crow" laws, applied to the American soldier, at times sharply decreased the efficiency of the unit . . . full discussion of the incidents cut the periods to a minimum and the men reacted intelligently to their situation at all times. Nevertheless the only effective solution was the complete removal of the squadron from the South.[118]

Such a strong indictment of local customs should be presented only as the personal opinion of one man on a highly controversial question. To present both sides, exactly the opposite point of view, in equally strong language (in connection with the same unit, the 456th Aviation Squadron at Daniel Field, Georgia), was contained in an "ASC Survey of Negro Organizations," which read as follows:

> The 456th Aviation Squadron was restricted because of laxity. In protest, all except one of the non-commissioned officers turned in their stripes. The men were reduced to privates and immediately transferred out. Morale has been favorable since the correction of these irregularities.[119]

It should be noted at this point that expressions of dissatisfaction with the training of Aviation Squadrons were not confined to the personal opinions of unit commanders and historians, or to the utterances of disgruntled enlisted personnel. From Headquarters, AAF, Brigadier General L. P. Whitten had written to the Commanding General, Air Service Command, Patterson Field, Fairfield, Ohio, as follows:

> 1. Attached are Inspector General's reports on the following Aviation Squadrons. . . .
> 2. Also attached is copy of memo for the Commanding General, Army Air Forces, from the Chief of Staff, which is self-explanatory.
> 3. It is apparent from the reports submitted that:

a. Completely inadequate supervision is being given to the proper training of these units.
b. Officers who are either weak or who lack the proper military background are being assigned command of these units.
c. Incompetent personnel are being assigned as fillers to units about to depart for overseas instead of such personnel being classified Section VIII [psychologically disturbed].
d. The type training being given these units is being improperly handled, as evidenced by lack of morale and discipline.[120]

The training of colored units of the other arms and services was not entirely satisfactory either. There was, for example, in July 1943 the incident of the fifteen hundred colored Ordnance troops at Kelly Field, San Antonio, Texas.[121] This was an extreme case, not at all typical of conditions generally, but it should nonetheless be recorded. The condition of these troops upon their arrival at that station, in the words of the Commanding General thereof, was a disgrace to the United States Army.[122] Not only were items of clothing and personal equipment in various stages of disrepair, but shoes were worn out and "several members of the organizations were barefooted, because their feet would not admit of shoes." [123] The physical condition of the men was equally poor. It was described by Brigadier General Paul C. Wilkins in an indorsement to the Commanding General, Air Service Command, as follows:

The physical condition of these troops upon their arrival was such that the Post Surgeon was directed to conduct an immediate physical examination of them all. This examination disclosed that 91 enlisted men were suffering from veneral [*sic*] disease, and 130 were unfit for general service, some of which [*sic*] were obvious without physical examination and should have been detected prior to their transfer to this Command, and the officers responsible for their transfer should be held financially liable for their transportation costs.[124]

The neglect from which the soldiers in these six Ordnance units had been suffering was almost unbelievable, and the lack of supervision and control evidenced by their condition seemed to General Wilkins to indicate that they must have been expected to take care of themselves. Vigorous measures were taken by the San Antonio ASC, therefore, to bring them into line. General Wilkins, assisted by various members of his area staff, took personal command. The troops were moved from the residential area of the post to South Kelly Field, "not . . . as a result of minor incidences [*sic*], but because of the fact that accommodations for 1,100 troops were requested, but approximately 1,500 were transferred;

which were too many for the area selected to accommodate." [125] Inspections were initiated; schools were set up for officers and non-coms; and all officers assigned, including instructors, were restricted to the post until they could set up a satisfactory training program. In addition, unfortunately, "in order to enforce discipline it was necessary to confine an excessive number of these EM to the Guard House, which was disparaging to the Courts-Martial Record of this station." [126] Nevertheless, conditions seemed to warrant drastic action, for at the time of General Wilkins' original inspection shortly after the men arrived it had been found that:

(1) Troops had had practically no training as soldiers.

(2) Officers, mess sergeants, and other non-coms were generally inexperienced and incompetent to command.

(3) No officers, at the time of the inspection, either instructors or company officers, were present, but were living in town and few had visited their troops. [The responsible officers were white.] [127]

Other situations were encountered in the training of such colored units as Quartermaster Truck Companies. Since Service Groups ordinarily contained two of these companies, the question arose as to whether or not one of the two could be colored and the other white. It was determined as a matter of policy that these two companies should be "assigned to each Service Group all white or all colored, but not mixed." [128] A similar policy, incidentally, was applied to Guard and MP Squadrons.[129] Another question arose as to whether or not colored personnel should be utilized in parent groups. In May 1943 the Acting Chief of the Military Training Branch, Headquarters, ASC, recommended that all Parent Service Groups be authorized to have white truck companies. "If this authorization is possible," he wrote, "it would eliminate a certain acute situation which will, beyond a doubt, present itself in case colored troops are used as instructors for white troops. If it is impossible to authorize white Truck Companies for Parent Service Groups, then every effort should be made to get at least one truck company of white soldiers attached to each Service Group in order to avert unpleasant situations that will arise." [130] This recommendation could not be followed, however; the mixing of colored and white truck companies was contrary to high policy. Therefore, in the training structure established 30 December 1943 one of the Parent Service Groups, the 40th, was authorized two colored Quartermaster Truck Companies. The rest

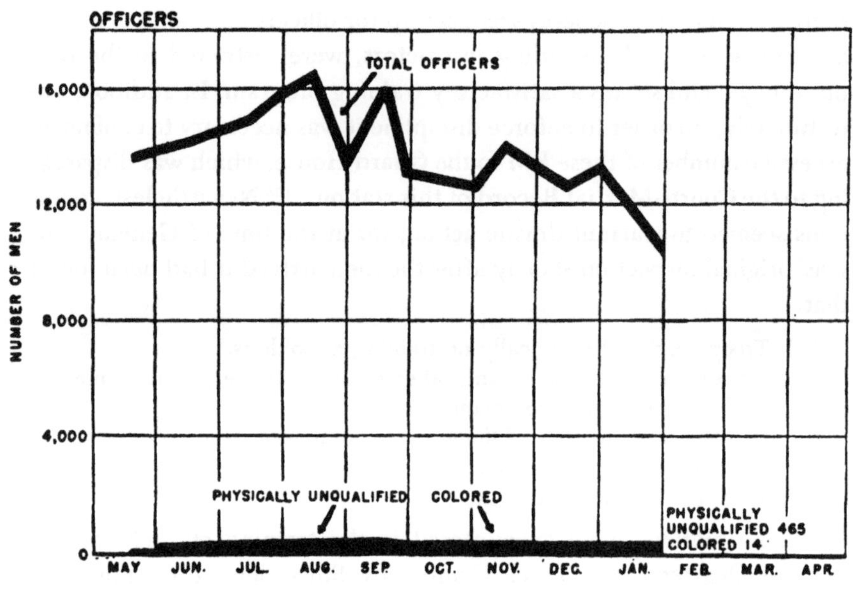


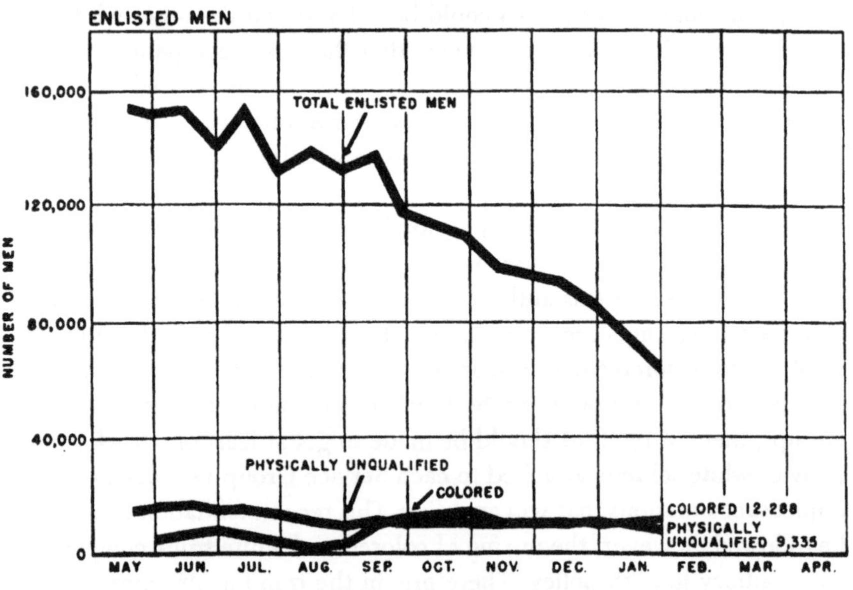


FIGURE VIII

MILITARY PERSONNEL, MINORITY GROUPS
Control Room chart, Hq ASC

were exclusively white. The other colored training structure unit authorized at that time was 703rd Chemical Maintenance Company (Avn).

To distinguish between colored and white units, the term (Separate) or (Sep) was used. Separate meant colored.* The policy of the Air Service Command, as expressed by its Commanding General, Major General Walter H. Frank (who was in command from 19 November 1942 [131] until 17 July 1945) [132] was not to mingle the two more than was done in other commands.[133] Furthermore, the Air Service Command refrained from activating any more tactical organizations for colored personnel than it had to. In January 1944 the excess of colored personnel on hand, not formed into tactical organizations, was so great that the Weekly Activity Report stated:

> To utilize excess negro personnel assigned to the Air Service Command, this Command has been advised that the casual shipment of 2000 EM, negro, will be prepared for overseas immediately, and new organizations will be activated to absorb excess ASC negro personnel.[134]

So far as commissioned personnel were concerned, in spite of the directives that colored troop units were to have colored officers wherever possible,[135] the evidence indicates that colored personnel were seldom used as commissioned officers within the Air (Technical) Service Command. Detailed statistics on this subject were contained in a chart in the Control Room, Headquarters, ASC, in which complete strength figures were given, especially for colored troops and troops physically unqualified—for overseas duty. (See Fig. VIII.) While the ratio of officers to enlisted men for white ASC personnel at the end of January 1944 was approximately one to six, that for colored personnel was only a little better than one to one thousand.[136] In fact, on 31 January 1944 there were only fourteen colored officers in the entire Air Service Command.[137] While there were 12,288 colored enlisted personnel in the command, as against approximately 60,000 whites, there were only these fourteen colored officers, as against considerably better than 10,000 whites.[138]

* *Separate* had other meanings also; for example, it was used to designate independent squadrons, split off from groups.

# ★★★ 10 ★★★

# STEPS TOWARD A SEPARATE AIR FORCE

## Planning for Integration

At headquarters, AAF, the presence of arms and services within the Army Air Forces, especially in tactical units, was regarded as a carry-over from the days when the air force was subordinate to the ground force and needed its help for every possible service. Air Corps officers were convinced that the existing structure was fundamentally unsound. Units and personnel designated as Quartermaster, Ordnance, and Signal Corps were part of the Army Air Forces and at the same time were identified with the technical branches of the Army Service Forces. In other words, they were expected to serve two masters. Overlapping efforts involved the technical branches in a number of intrinsically Air Corps activities. Ordnance functions, for example, duplicated aircraft armament; Signal functions duplicated air communications; and the supply activities of all arms and services were similar to AAF supply. These inconsistencies spurred the integration program, which steadily gained momentum from 1943 to 1945.

The underlying concept (in addition to the idea of a separate air force) was the basic principle that like activities and functions should be grouped together to assure a natural flow from raw material to end-product, whether the end-product be a complete aircraft, a trained airman, or a tactical service organization. In short, the approach was functional. Experience in the theaters had demonstrated the futility of converting or distorting ground force units to air force uses.[1] The logical grouping of functions for the accomplishment of Air Corps missions was the following:

(1) Personnel and Administration
(2) Supply and Maintenance
(3) Training and Operations

The first group included the care of personnel as individuals. Conflicting procedures and duplications in clerical staffs and personnel records would be eliminated by consolidating these functions. The second included the providing and maintaining of equipment and real property. The third, the utilization of the end-products of the other two, personnel and materiel, to perform the tactical mission of the organization concerned. Careful analysis of the arms and services indicated that all functions would fall into one of these three groupings. Similar activities would be combined and duplications would be avoided.[2]

Obstacles to the program existed both within the Army Air Forces and without. The principal internal deterrent was that prior to 1 January 1944 there were two organizations on each base responsible for aircraft services: base services, which performed limited functions under the base commander, and the sub-depot, answerable to the Air Service Command. The other branches were quick to notice that the Army Air Forces had not consolidated what were, in effect, its own arms and services, and they used this condition as an argument against integration. Moreover, special staff officers (ASWAAF's) were reluctant to adopt the concept of integration for fear they would lose access to the commanders on whose staffs they served, and thereby compromise their positions with the chiefs of the technical services.

The technical services of the Army Service Forces objected to the program because they felt that they should retain the identity and integrity of their respective departments in all elements of the army. In view of the long and successful history of the branches, they felt that they were better prepared to serve the Army Air Forces than the latter was to serve itself. AAF officers replied that they had no desire to take over the development and procurement of items common to the army as a whole, but merely to obtain the personnel and units currently assigned to the Army Air Forces and to make them integral parts of the command. This involved drastic innovations, however, and AAF membership on the General Staff was always well below the 50 percent * contemplated in the War Department reorganization of March 1942.[3] Each succeeding conference on the subject was characterized by elaborate explanations,

* It was felt by many that the numerical weakness of the AAF within the General Staff was caused by the lack of a sufficient number of qualified Air Corps officers, and not by discrimination against the air forces.

for the benefit of ground force officers, of the organization and mission of the several parts of the Army Air Forces.[4]

Another obstacle was the fact that at domestic air bases the Army Service Forces retained control over a number of operations historically associated with its technical branches. More specifically, airdrome maintenance, repairs, and utilities were identified with the Corps of Engineers; field communications with the Signal Corps; laundries, bakeries, and salvage with the Quartermaster Corps; vehicle and weapon repair with the Ordnance Department. Special staff officers, and to some extent the base commanders themselves, were thus responsible both to the Air Corps and to the ASF Service Commands.

The measures taken by the Army Air Forces to overcome these obstacles extended over a two-year period, beginning in the summer of 1943. They resulted eventually in a large degree of independence for the air arm. The basic idea was presented to the General Staff in a series of studies pointing out the inconsistencies of the current arms and services system, the savings of manpower possible with integrated organization, and the benefits of standardization.[5] The first study was presented to Lieutenant General Joseph T. McNarney, Deputy Chief of Staff, over the signature of General Arnold in July 1943.[6] This study recommended the elimination of branch distinctions, and the detailing to the Air Corps of all personnel assigned to the Army Air Forces in this country and overseas. On 17 August 1943, before an answer was received, the three major commands were directed to make a joint study of the organization of the army for the purpose of eliminating duplications and consolidating similar supply and maintenance activities.[7]

To prepare the AAF portion of this report the Army Air Forces created a board headed by Major General Delmar H. Dunton, which convened at the School of Applied Tactics in Orlando, Florida.[8] The board recommended the reorganization of Air Depot and Service Groups, and again the elimination of branch distinctions. The joint report of the three major commands was submitted to the Deputy Chief of Staff on 10 October 1943,[9] and the AAF proposals were approved on 20 October 1943.[10] In this manner the idea of integration was established.

Upon approval of its recommendations the Army Air Forces immediately started conversion, the first step being the appointment of an Arm and Service Integration Committee in Washington under the Chief of Management Control.[11] On 3 December 1943, however, before any action could be taken by this committee, a proviso was added to the War Department's unconditional approval of 20 October 1943, requiring a 20 percent saving of personnel in the integration of all overseas units, and temporarily suspending approval of the integration of zone of interior activities.[12] Doubtless, these rulings had been caused by the representations of the Army Service Forces, but the Army Air Forces felt secure in its position, and continued to base its actions on the hope of ultimate independence.

The conflict of command in AAF activities at domestic air bases was eliminated when the sub-depots were transferred to the air forces and commands on 1 January 1944.[13] The Air Service Command resisted this change. In fact, even after it was accomplished, many of the command's actions were premised on the possibility that the sub-depots would be returned.[14] The change was made, nevertheless, and the Army Air Forces was thereafter protected from the perennial efforts of the Army Service Forces to gain control over all air bases.[15] Other internal reforms accomplished during 1944 were the establishment of a standard functional plan for domestic air base organizations, preparing the AAF for control over technical base services,[16] and the acquisition of budgetary control over the necessary funds.[17]

Meanwhile, the Arm and Service Integration Committee was laying the groundwork for the absorption of associated activities, and planning a new logistical system for tactical units. Five sub-committees were set up to consider the following aspects of integration:

Removal of branch insignia
Classification of AAF personnel
T/O's for service units
T/O's for combat units
T/E's [18]

Approval of the principles set forth so often by the Army Air Forces was finally obtained on 28 December 1944, when the Deputy Chief of Staff issued a guide for the relationship of the three major commands entitled "Service Responsibilities," [19] which in effect overruled the objections of the Army Service Forces. Shortly afterwards regulations were promulgated for the transfer of arm and service personnel to the Air Corps,[20] and at this point the integration of personnel was declared publicly to be an accomplished fact. Under the auspices of the Integration Committee, a revised classification manual was started to standardize procedures for all AAF personnel.

Changes in Tables of Organization and Equipment (T/O&E's) for tactical units, however, as noted earlier, resulted not only from this drive for integration, but also from the experience of the theaters of operations.[21] Where landing fields were scarce or combat units numerous, it had often been necessary to place two or more groups at the same station. Thus, Service Groups were put side by side with combat groups, and the command of the two was often combined. Although the China-Burma-India Theater (where the units were scarce) had retained the separation of functions established by AAF Regulation No. 65–1, most theater commanders had found it necessary on occasion to place Service Groups and base functions under the combat group commanders, thereby altering the structure established by the War Department.[22] The basic organizational problem in AAF logistics during 1944 and 1945, therefore, was to establish integrated, functional teams based on the current needs of the air arm, and not on the historical structure of the army, as the army had been organized prior to the emergence of air power.

## The Service Group (Special)

The activation of the first Service Groups (Special)—also called Service Groups (New Type)—coincided almost exactly with the introduction of the Very Heavy Bombardment (VHB) Program of the B-29 "Superfortresses." The original object of the new style of organization in these units had been to accomplish the integration of arms and services, with the addition of conventional base functions. But the whole project was so closely associated with the preparation of VHB units for the global Twentieth Air Force that the two programs were quickly identified with one another.[23] Furthermore, they were both under the jurisdiction of the same special body, the Executive Committee of the B-29 Liaison Committee, headed by Brigadier General K. B. Wolfe. The significance of this arrangement was that basic decisions affecting both programs were not at first made by the Air Service Command,[24] so that plans for the supply and maintenance of B-29's were not consistent with conventional ASC methods, and it was necessary for this command to "sell" the idea of Air Depot and Service Groups to the Air Staff all over again.

The notion of the B-29 Liaison Committee seems to have been that the Superfortresses were large enough to be self-sufficient, and that they could "live off the land." [25] Normal operational air bases similar to those employed for other aircraft, the committee felt, could be dispensed with if the crew members were given special training in first and second echelon maintenance, and if the necessary tools were carried along in specially constructed tool-cribs. Shuttling back and forth between continents, these long-range planes could return to the States whenever top echelon work was required.

This imaginative concept of the B-29's role in aerial warfare proved to be too sanguine; actually B-29's demanded more attention than any other AAF plane. The sheer size of the B-29 (141′ wingspread as compared with 104′ for the B-17) created a need for larger tools and more specialized maintenance equipment than any previously used. The new features, such as central station fire control, pressurized cabins, and a multitude of electrically driven accessories and new radar devices caused further complications in overseas maintenance.

Nevertheless, the first organization for B-29's, the 58th Bombardment Wing, was to have no Service Groups or Air Depot Groups at all. Although the Air Service Command anticipated that Headquarters, AAF, would soon direct that these groups were to be equipped and trained, no such directive was issued. Instead, General K. B. Wolfe in May 1943 contemplated a closely knit combat force, quite independent of service and maintenance organizations, and this plan remained the official doctrine until late fall.[26] In October 1943 General Wolfe's program was modified only by the proposal that each Bombardment Squadron of the 58th Wing be accompanied by a Maintenance Squadron for first and second echelon work. Meanwhile, in spite of this apparent indifference to service units, Brigadier General E. E. Adler of the Air Service Command raised the question of the organization and training of support groups for what was then known as the PQ Project at an Air Forces Conference, held in Colorado Springs from 20 to 22 September 1943.[27] Nothing came of the suggestion then, and the "live off the land" theory was again cited as the reason for inactivity. It was not until 12 November 1943 that a directive was issued by Headquarters, AAF, requiring the preparation of support units for B-29's.[28] At that time General Adler was informed over the telephone by Colonel M. A. Kelly, Office of the Assistant Chief, Air Staff, Training, that two Service Groups and one Air

Depot Group would be prepared immediately to move overseas with the 58th Wing.

The two Service Groups designated were the 25th and the 28th Service Groups, both old-style organizations which had been acting as parent groups and were already in condition to go.[29] The 22nd Air Depot Group, also an experienced unit, was likewise designated to work with B-29's. These groups were prepared for immediate shipment, with great secrecy, and all personnel except a few specialists arrived in India on 30 April 1944. There they were reorganized to form four Service Groups (Special), the 25th, 28th, 86th, and 87th. The specialists included certain Signal Corps detachments, which were trained at Warner Robins, and central station fire control experts, trained at Lowry Field, Colorado, and at the Oklahoma City ASC.[30] These men remained behind for technical training, and then flew to the theater with the combat echelons.

In Service Groups (Special) the personnel of the other arms and services were to be absorbed into three streamlined squadrons: a Headquarters and Base Services Squadron, a Materiel Squadron, and an Engineering Squadron.[31] The special staff organizations of the older groups were thereby to be eliminated. Each Service Group (Special) was to serve one combat group, and to operate a base complete with Finance service, fixed communications, Medical dispensaries, interior guard, internal security, utilities, firefighting, and motor transportation.[32] Moreover, if tactical situations necessitated it, they were to supply combat groups at dispersed airdromes through refilling and distribution points, and to keep them in commission by Mobile Repair Units. Normally, they would be backed by a centrally located AAF Depot which would perform fourth echelon services for several groups.

One of the chief advantages of the new organization was that there were to be fewer elements of command (three instead of nine squadrons), and if the personnel of one unit were not busy on a particular day, they could be used to help another unit with a heavier workload.[33] Formerly the prerogatives of each of the arms and services had been guarded so jealously that it was rare for Quartermaster personnel, for example, to pitch in and help Ordnance to complete a rush project, regardless of its priority. Now all supply personnel (in the Materiel Squadron) and all maintenance personnel (in the Engineering Squadron) could be told where to work and what to do without consulting officers of the other arms and services.[34]

In January 1944 the needs of the XX and XXI Bomber Commands were established at twenty-four support organizations: sixteen Service Groups and eight Air Depot Groups. All of the Service Groups were activated in accordance with the new T/O's except two, the 303rd and the 330th, which were old type units reorganized while still in this country to form four Service Groups (Special). Thereafter the identification of the Service Group (Special) with the B-29 program was complete.[35]

Obtaining personnel for the twenty-four new support groups, however, was almost impossible, for this requirement was over and above the Army Air Forces' 273 Group Program.[36] Moreover, the demand arose early in 1944 when the Army Ground Forces was pressing the Army Air Forces for approximately 157,000 men, needed to activate fifteen Infantry divisions. Significantly, the War Department recognized the high priority of the AAF requirement, and the AGF demand for fifteen additional divisions was sacrificed in favor of Air Depot and Service Groups and other AAF units. The extreme importance of support groups for the VHB Program was thus belatedly admitted.* Once the decision had been made to retain the personnel demanded by the Infantry within the Army Air Forces, the requirements for VHB service units were filled by the Technical Training Command and the Air Service Command, with help from the Caribbean Defense Command and certain inactive theaters.[37]

Training for Service Groups (Special) was based on a six months course which began when each group reached 90 percent of its organizational strength.[38] Unit training lasted approximately sixty days, and was carried on at Robins Field, Georgia; the ASC Training Center, Fresno, California; Kelly Field, San Antonio, Texas; and Tinker Field, Oklahoma City, Oklahoma.[39] Following this shakedown period, the groups were moved to designated bases in the Oklahoma City area for combined and final phase POM training. In this area they were based at one of four Kansas air fields: Smoky Hill, Great Bend, Walker, and Pratt Army Air Bases. Here they received OTU training with the Bom-

* It should be remembered in this connection that the Ninety Division Program of the Army Ground Forces had left but a small margin of safety. The grand plan was "figured close"; at the end of the war all existing divisions, with one or two exceptions, had been committed, and there was nothing remaining for emergencies.

bardment Groups of the 58th Wing, then in transition training. Later, as the B-29 program expanded, four additional bases were opened in Nebraska: McCook, Grand Island, Harvard, and Fairmont. After the departure of the 58th Wing, these bases came under the control of the Second Air Force, which was charged with the training of the 73rd and succeeding Bombardment Wings of the XX and XXI Bomber Commands.[40] The national training structure in July 1944 was composed of the units listed in Table 13.[41]

TABLE 13

TRAINING STRUCTURE, JULY 1944

4500th AAF Base Unit (Serv Gp), Venice, Florida
4501st AAF Base Unit (Serv Gp), Avon Park and Lakeland, Florida
4502nd AAF Base Unit (Serv Gp Sp), Fresno, California
4503rd AAF Base Unit (Serv Gp Sp), Tinker Field, Oklahoma
4504th AAF Base Unit (Serv Gp Sp), Tinker Field, Oklahoma
4505th AAF Base Unit (ADG), Kelly Field, Texas
4506th AAF Base Unit (Serv Gp Sp), Fresno, California

The first Air Service Groups (Special) * entering the Oklahoma City area did not go directly to the training bases, but reported to Tinker Field for a month or so to accomplish processing, including immunizations and furloughs, and basic military training.[42] Thus, a pool was established at the beginning of the program to insure an uninterrupted flow of training at the Kansas and Nebraska airfields.[43] The 4514th AAF Base Unit (Air Service Squadron, Training) was established at Tinker Field to assist in these tasks. The trainee Service Groups operated their own headquarters, mess halls, motor pools, and supply warehouses; OTU equipment was issued by a special "custodial unit" of the Supply Division of the Oklahoma City ATSC. Under this arrangement branch equipment was issued in "sets" for individual groups, so they could get their equipment from a single agency instead of a number of sources.

After the processing period the early groups went by motor convoy from Tinker Field to the training bases, bivouacking for several days en route. Later groups went directly from Fresno and Warner Robins. At the bases they served under small training teams, consisting of fourteen

* The designation Service Group was changed to Air Service Group on 4 Nov 1944. (T/O&E 1–452T, 4 Nov 1944.) The change was made because of misunderstandings overseas, where theater commanders in some cases assumed that Service Groups were labor battalions for use on general fatigue details.

officers and fifty-three enlisted men,[44] permanently assigned to supervise combined training. There were nine of these training teams, one for each base; their functions were similar to those of the old ASC parent groups—with greater emphasis on instruction and less on supply and maintenance. One of the main features of this period was the training of sections or shops, as distinguished from squadrons, capable of separate operations when split off from the rest of their units, as they often were overseas.[45] On several occasions, wrecks of B-29's in the vicinity gave the men opportunities for practical work, and they were able to fly back to the depots in ships which they had repaired.[46]

Inconveniences caused in the past by the flow of personnel in and out of training bases, and to and from special schools, were partially solved in this case by the establishment of a centralized B-29 School at Oklahoma City, Oklahoma, which combined courses ordinarily given at factory schools all over the country.[47] Although this school did not eliminate factory schools, it provided a total of eighteen courses including such subjects as propellers, central station fire control, C-1 automatic pilot, and numerous electrical instruments. Practical experience was obtained from live models such as actual central station fire control turrets, which were assigned for this purpose. From its opening in September 1944 until May 1945 the school trained more than 1,700 soldiers and 600 civilians for periods of from one to six weeks. Its value was confirmed by the attendance of factory representatives from more than thirty private corporations.

Air Depot Groups for the B-29 project were trained by the 4505th AAF Base Unit, a parent Air Depot Group, at Kelly Field, Texas.[48] These units were not reorganized, as were Air Service Groups, and there was no integration of fourth echelon arms and services except for the addition of an Ordnance Section to the Hq and Hq Squadron. To facilitate on-the-job training in the depot's shops and warehouses, a Training Liaison Office was established in September 1944 as part of the Military Training Section of the headquarters. The function of this office was the placement and supervision of group enlisted men for the common benefit of the trainee units and the Maintenance and Supply Divisions of the depot.

During final phase (or POM) training, the personnel of Air Service and Air Depot Groups returned from special schools, and detached service, and the groups assumed the composition they were expected to have on movement day. When warning orders were received, packing squads were dispatched from the depots to pack, crate, and mark all ASWAAF equipment and such AAF equipment as was not handled at Intransit Depots. Altogether, from November 1943 to May 1945 a total of twenty-four Air Service Groups (Special) and six Air Depot Groups were trained by the Air Technical Service Command.[49] As part of the same program there were also two Aircraft Repair Units (Floating) and two Aircraft Maintenance Units (Floating) which served as sea-based service facilities for the B-29's. Approximately 1,200 officers and 20,000 enlisted men were included in the units listed above.

A typical theater of operations in the early months of 1945 contained both old type and new type Service Groups in simultaneous operation. (See Fig. IX.) The former serviced two combat groups each; the latter only one. Separate Aviation Gas and Oil Depots, and Quartermaster, Medical, Engineer, Chemical, Signal, and Ordnance Depots were still in existence, although ASWAAF units within Service Groups (Special) had all been integrated. The Air Force General Depot, with one or more Air Depot Groups, was still organized in the conventional manner.

## Later Developments

In May 1945, Headquarters, AAF, directed that the training of Air Service Groups and all allied third echelon units be transferred from the Air Technical Service Command to the Continental Air Force (which included the four numbered continental air forces of the past), while Air Depot Groups, and associated fourth echelon ASWAAF units, were to remain with the Air Technical Service Command.[50] The reason for changing was not entirely clear, and of course the ATSC objected to the loss of its Air Service Groups. The Continental Air Force (CAF) was reluctant to assume the added responsibility, as it was inadequately staffed and equipped to train these units.[51] The CAF protested that the new arrangement was based, in part at least, on the incorrect theory that redeployed units would require little training, whereas actually in most cases they would have to be 70 percent remanned. The CAF therefore recommended that it be allowed to leave unit training with the ATSC for the time being, but that it pick up newly activated or redeployed units for combined training. If the change were made, however, the nine ATSC training teams of the 4514th AAF Base Unit,

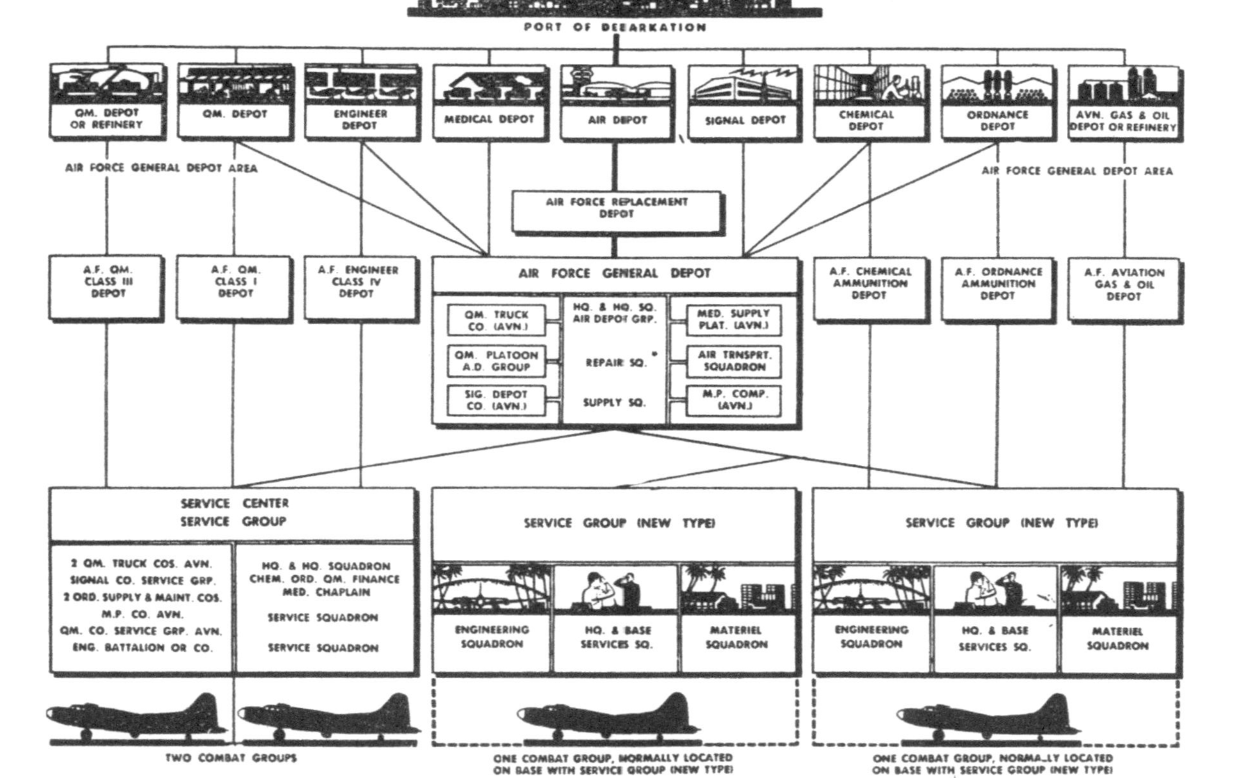


FIGURE IX

AIR FORCE SUPPLY IN A THEATER OF OPERATIONS

Tinker Field, Oklahoma, should be frozen in their present duties, and directed to assist the air forces in every way possible.[52]

The best explanation for the decision of Headquarters, AAF, was the fact that two distinct trends of reorganization had manifested themselves in the theaters. One was the integration of ASWAAF units into the Army Air Forces; the other was the parallel movement within the air forces toward the integration of combat units and service units under the "deputy plan." [53] It was the latter which caused the transfer of third echelon service units to the air forces (whence they had come at the beginning of the war), for only if combat units and service units were trained by the same command could they be integrated into effective combat-service teams. Once this aim was accomplished, all units at the same stations could be commanded by the combat group commanders, while Air Service Group commanders could act as Deputies for Administration in charge of maintenance, supply, and base services.[54]

Integration under the deputy plan had been initiated with the publication of the T-Type Tables of Organization for Service Groups (Special) late in 1944; it was amplified early in 1945 by the more advanced R-Type Tables, which included even Air Depot Groups. The main difference between the T-Type Tables and the R-Type Tables was that certain administrative, supply, and special maintenance personnel, assigned to the combat groups under the "T" system, were added to the Service Groups under the "R" system. The transfer was accomplished by attaching "type" columns, which corresponded to the particular type of combat groups being serviced, to the basic Air Service Group squadrons.[55]

Meanwhile, as a result of the hesitation of the Continental Air Force, Headquarters, AAF, wavered several times before making its decision definite. Finally, after a conference in Washington, June 1945, it was determined that the nine training teams of the 4514th AAF Base Unit at Tinker Field, Oklahoma, together with three additional training teams of the 4510th at Robins Field, Georgia, would be transferred bodily to the Continental Air Force.[56] Thus, responsibility for the training of Air Service Groups was delegated to the Continental Air Force without restriction. The jurisdictional differences between the two commands were settled by transferring the training structure units along with the trainee units. Air Depot Groups remained with the Air Technical Service Command.

The problems anticipated by the Continental Air Force never became serious, happily, because few of the units scheduled for redeployment actually had to be retrained and shipped to the Pacific. When the war in the Pacific ended, sooner than expected, the training program of the AAF quickly folded up. Directives effective V-J Day provided that only the units connected with the VHB program should be shipped overseas as planned, and that these should be shipped only if they were already at or en route to the Ports of Embarkation.[57] If additional personnel were needed, training structure personnel were to be used, provided they were qualified for overseas service, were not over 38 or under 19 years of age, and had point scores for separation from the service of less than 70.[58]

The demobilization of the army after the Japanese surrender was exceedingly rapid, and the training structure of the Air Technical Service Command was correspondingly reduced. The ATSC itself was sharply diminished, and plans were made for the continuance of only seven of the eleven control depots. The others were to be used as satellite supply and maintenance depots for the storage of surplus materiel. Eventually, all ATSC training activities were consolidated in the San Antonio area under the jurisdiction of the 4505th AAF Base Training Unit (Air Depot Group) at Kelly Field, Texas. The responsibility of the command (now renamed the Air Materiel Command) for the post-war Interim Air Force was limited to "the activation, organization, and training of assigned service organizations on the depot level, on-the-job training peculiar to the requirements and missions of these organizations, and such other specialized aircraft and communications maintenance training of individuals as [was] required in cases where resources for such training [were] not available in other commands." Air Depot Groups in turn were subjected to the process of integration; under the R-Type Tables of Organization they were redesignated AAF Air Depots, and consisted of the following units: Hq and Base Services Squadrons, Air Depot; Air Supply Squadrons; Air Repair Squadrons; Air Vehicle Repair Squadrons; Air Ammunition Squadrons; and Motor Transport Squadrons, Air Depot. The development of the Interim Air Force into the 55 Group Program, and subsequently into the 70 Group Program, is current news.

★★★★★

# CONCLUSION

IN SUMMARY, it should be repeated that the air service unit program consisted of two major, overlapping operations: the development of new units, and the establishment of a national training structure to mass-produce those units. At first the staff officers of the Air Service Command were preoccupied with such questions as the composition or functions of new groups, and the sources of personnel, facilities, or equipment. In this phase they participated in maneuvers and service tests, to determine the efficiency of particular organizations and work out procedures which could later be standardized. As they assumed increasing responsibilities, they turned to a consideration of bulk allotments, or the parent group system. Their interest at this point was to introduce refinements such as the centralization of personnel replacement depots or the establishment of an automatic system of supply. By the fall of 1943 the mass production of service units was beginning to be efficient.

Then, suddenly, the whole program was interrupted. Service units were needed for the invasion of Europe immediately; there was no time to mass-produce them. By the terms of the Bradley Plan approximately half the training program was transferred bodily to the British Isles—the men unorganized and their equipment distributed by guesswork. For those in authority it was a hard decision, like many others in the war, but the alternative was to postpone D-Day, perhaps for a year. The unscrambling of these hasty shipments and the organization and training of the Eighth and Ninth Air Forces in the United Kingdom were considered by many to have been one of the major miracles of the war. That, however, is not within the scope of this study. So far as the domestic program was concerned, the procedures had been worked out, and the training structure was now free to operate sedately on a much reduced basis.

As it happened, there were many demands for new special projects, and the training structure was kept busier than anyone had expected. Most of the demands were for the war in the Pacific. A striking occurrence in this period was the successful combination of the VHB service unit program for B-29's with the integration of arms and services, and both of these programs owed much of their success to the diminished pressure for D-Day troops. When the War Department decided on the Bradley Plan, it not only filled an ETO need, but also it enabled the domestic training structure to prepare with greater efficiency for the final effort in the Pacific, and at the same time to solve, or attempt to solve, the organizational problems which had bothered the air arm from the beginning of its history.

What of present value can be learned from the preceding chapters? In a few years, scientific innovations will make the successes or failures of military organizations so different from those here described that these conclusions will be obsolete. But, for today, the trends which this summary emphasizes may still have a bearing on military events.

For instance, during World War II the development of new aircraft and new technical equipment did not eliminate the necessity for establishing supply and maintenance organizations in overseas theaters. The elimination, or the drastic reduction, of these organizations was often suggested, and several times it was proposed that new combat aircraft could "live off the land" or be flown back to the United States for overhaul and repairs. But after careful consideration it was found that these arrangements would not be satisfactory. The effectiveness of the air arm depended upon mobility, and mobility depended upon a network of dispersed repair and supply bases in every theater of war. The bigger the planes, the harder they were to service. Their constantly increasing range did not seem to alter the need for overseas depots. It was an unusual aircraft that could be flown across the ocean with instruments that didn't work, or defended with armament that couldn't fire.

Not once, but many times, it was necessary to revert to the standard procedures of supply, maintenance, and reclamation which had been worked out in maneuvers at the beginning of the war. The experience of those who serviced newly developed equipment, such as jet-propelled fighters, was similar in this respect to the experience of others engaged in special projects, such as the organization of Floating Maintenance Units for the Pacific, or Depot Units (Army) for Ground Force aircraft in Italy.

Granted that some service units were overstaffed, and even that a few were later considered to have been altogether useless. Yet, before demanding the elimination of these organizations, one might pause to reflect that in the crisis of combat the War Department was willing to accept new AAF units, including a great number of service organizations, as a substitute for fifteen Infantry divisions.

In any case, the decision to keep or to alter AAF service organizations was never a simple question to be decided on the basis of one or two factors. Each new project had to be considered in the light of the whole system of AAF logistics, and that system had to be adjusted to meet the political and military requirements of the various theaters, and the exigencies of naval and submarine warfare. The training doctrine for service and maintenance organizations thus had to be dependent upon events in the theaters of operations, and upon the procedures and capacities of Industry. The ATSC (AMC) was never able to control these factors.

The flying officers who have recently (1948) been placed in charge of the training of air service units have been put in a curious position by the policy of permitting only rated flyers to serve on active duty. Most of them unconnected with maintenance and supply since Pearl Harbor, these flyers must now perform the administrative tasks from which, in the war years, they asked to be released. In establishing the new doctrine for tactics and strategy, it will not be surprising if they try once again to eliminate service organizations, and to add these functions to the combat units of their own experience—the same procedure which so hampered the air forces in the early months of World War II.

Undoubtedly there is a place for a few flying officers in service and maintenance organizations. But if they are to be given a monopoly on positions of authority, and if as a result of this policy the experience of those who developed the service and maintenance organizations of World War II is to be lost, it is hard to see how any permanent advantage to the United States Air Force is to be obtained. Favoritism to flying officers, or to Regular Army officers (or, some would add, to officers) is inconsistent with the technical mission of Air Materiel Command organizations.

Another point to be emphasized is the fact that the trend toward centralization, which has been so marked in this air service unit program, (always excepting the Bradley Plan decision to do a lot of the job in England) has had unforeseen consequences. The trend itself has been technological—the result of changes in methods of communications and in the physical techniques of doing things. It is a characteristic of big organizations, and can be seen in the political, military, and industrial arrangements of many nations.

The obvious effect of this policy of centralization has been to increase the military vulnerability of the nerve centers from which control is exercised over the activities of large commands. There is also what might be called the moral vulnerability of these same nerve centers. As these chapters have shown, efficiency is often increased when procedures are standardized, and not left to the discretion of subordinates. In the process of standardization, however, a number of social matters (housing and recreation, for example) are also directed by the central authorities, and here the benefits of centralization are sometimes uncertain. Traditional patterns of behavior, which were once somewhat fluid and subject to occasional variations, are solidified consciously in regulations, which permit no deviations. In this situation segregation, or discrimination of any sort, can no longer remain a matter to be defended as a local custom or an individual decision; it becomes a command policy and does not have to be defended at all.

No study of administration on a federal scale would be complete if it failed to point out this dilemma. While no one would deny that all agencies of the Federal Government ought to respect local customs, it should be understood that it is increasingly hard for them to do so. In the Southern states, this respect for local customs is prompted not only by a feeling of obligation to the states, but also by the practical advantages in concentrating military training in the South, where it is cheaper to feed, clothe, house, and care for the troops. But it is also true that the local customs of other sections of the country have to be respected, and that any rigid system of segregation based on race or color is apt to create as many administrative difficulties in some areas as it might eliminate in others. This discussion is an attempt to show that a properly unified policy moves on from technological theory to the intricate subject of social action.

★★★★★

# REFERENCES


## INTRODUCTION

1. See speech by Brig Gen E. E. Adler in "Proceedings—ASC Theater Supply Conference, Patterson Field, Fairfield, Ohio," 1–4 Nov 1943.
2. "Development of the Air Force Logistical System," a memorandum prepared in 1943 by Col D. R. Stinson, Ass't Chief, Supply and Services Division, AC/AS, MM&D (hereafter cited by title only).
3. AAF Reg. No. 65–1, 14 Aug 1942.

## CHAPTER 1

1. AAF Historical Study No. 25, pp. 2–3.
2. *Encyclopaedia Britannica,* 15th Edition (1947), Vol. I, "Air Forces."
3. AAF Historical Study No. 10, p. 2.
4. *Encyclopaedia Britannica,* 15th Edition (1947), Vol. I, "Air Forces."
5. Public Law No. 242, 41 *Stat.* 759.
6. *Annual Report,* Chief of Air Service, 1923, p. 1.
7. AAF Historical Study No. 10, pp. 2–3.
8. AAG 320.2 (12–19–34) Misc. (Ret)-C, 31 Dec 1934.
9. AR 95–5, 20 Jun 1941.
10. AAF Historical Study No. 10, pp. 23–26. See also *The Army Air Forces in World War II,* Vol. I, pp. 257–267.
11. WD Cir. 59, 2 Mar 1942.
12. *Ibid.*
13. Interview with Lt Col Vincent T. Cannon, 10 Mar 1945.
14. "Regular Air Corps Active Organizations, Continental U.S. Only."
15. Air Service and Air Corps Cirs. No. 65–11.
16. Identical wording was used in several Air Corps Cirs.: 65–11A, 29 Sep 1930; 65–11B, 5 Dec 1930; 65–11C, 13 Jan 1932.
17. Maj H. H. Arnold, Chief, FSS, to CO, Selfridge Field, 19 Jun 1930, in TSAGD 352.1–4 (Maintenance and Operation of Aircraft).
18. ATSC Historical Monograph, "A History of the Maintenance of Army Aircraft in the Continental United States, Part I, 1921–1939," Chapter II.
19. AAG 320.2 (12–19–34) Misc. (Ret.)-C, 31 Dec 1934.
20. AAF Historical Study No. 10, Chapter I.
21. TAG 320.2 (5–5–36) Misc. (Ret)-MC, 8 May 1936.
22. Sec 102, Public Law 781, 76th Congress, 9 Sep 1940.

23. ATSC Historical Monograph, "The Air Service Command, an Administrative History, 1921–1943," pp. 23–25.

24. Exempted from control of the Corps Areas by AR 170–10.

25. Air Service (later Air Corps) Cirs. No. 65–11.

26. Sec IV, "Supply," Air Service Cir. No. 65–11, 9 Jan 1926, and Air Corps Cir. No. 65–11, 5 Apr 1929.

27. TAG 320.2 (6–19–40) M (Ret), M-C, 22 Aug 1940, effective 1 Sep 1940.

28. "Development of the Air Force Logistical System," p. 1.

29. Col David H. Baker to CG, 8th Air Force, "AAF Reg. No. 65–1," 25 Apr 1943, in the Supporting Documents of the "History of the Eighth Air Force Service Command."

30. Maj Gen George H. Brett to CG's, GHQ Air Force, The Training Centers, Air Corps Technical Schools, Chief, Materiel Division, and Col H. J. F. Miller, "Organization of Provisional Air Corps Maintenance Command," 14 Feb 1941, in Supporting Documents of "The Air Service Command, an Administrative History, 1921–1943."

31. TAG 320.2 Air Corps (4–25–41) M (Ret) M, 24 Apr 1941.

32. Same as No. 30, above.

33. AAF Reg. No. 65–6, 7 Feb 1942.

34. See dates in "Complete List of Service Groups," Appendix B, Supporting Documents.

35. "Development of the Air Force Logistical System," p. 3.

36. Maj Gen George H. Brett, Chief of Air Corps, to CG's, Southeast, Gulf Coast, and West Coast Air Corps Training Centers, the Air Corps Technical Training Command, and all Stations under the direct control of the Chief of Air Corps, "Activities and Operation of the Air Corps Maintenance Command," 28 Aug 1941, in TSHIS-2 files.

37. TAG 320.2 (12–8–41) MR-M-AAF, "Constitution and Activation of certain Air Corps Units," 27 Dec 1941, and AAF Reg. No. 65–7, 7 Feb 1942. Also Col William W. Dick to CG's of the Air Forces and Commands, "Army Air Force Sub-depots," 24 Jan 1942, in Supporting Documents of "The Air Service Command, an Administrative History, 1921–1943."

38. "Development of the Air Force Logistical System," p. 2.

39. AR 95–5, 20 Jun 1941.

40. AAF Reg. No. 20–4, 17 Oct 1941.

41. TAG 320.2 (12–9–41) MR-M-AAF, 11 Dec 1941, "Jurisdiction of the Air Service Command."

42. "Development of the Air Force Logistical System," p. 3.

43. Col William W. Dick, Air Adjutant General, to CG's, all Air Forces in Continental United States, and CG's, all Army Air Force Commands, "Responsibility for Service Units," 22 Jun 1942, in TSAGD 322 (Organizations—AAF Units).

44. "Development of the Air Force Logistical System," p. 3.

45. ASC General Order No. 29, 8 Apr 1942, and several subsequent General Orders.

46. Memo for General Miller by Brig Gen Clements McMullen, Chief, Overseas Division, ASC, 16 Aug 1942, in TSAGD 322 (Organization—AAF Units).

47. For further information see *The Army Air Forces in World War II*, Vol. I, pp. 294–5.

48. AAF Reg. No. 20–9, 14 May 1942.

49. "History of the Air Service Command in the European Theater."

50. "Air Force Administrative Organization and Procedure," 23 Jul 1941, in TSHIS-2 files.

51. *Report on the Army, July 1, 1939 to June 30, 1943*. Biennial Reports of General George C. Marshall, Chief of Staff of the United States Army, to the Secretary of War. (Hereafter cited as *Report on the Army*.)

52. Interview, Mr. Paul Angle with Brig Gen E. E. Adler, 10 Mar 1944.

53. *Report on the Army*, p. 90.

54. For further details see AAF Historical Study No. 29.

55. Interview, Mr. Paul Angle with Brig Gen E. E. Adler, 10 Mar 1944.

56. *Report on the Army*, p. 91.

## CHAPTER 2

1. Memo for the Chief of Air Corps by Brig Gen George H. Brett, Acting Chief of Air Corps, "Depot Supply and Repair Squadrons for Overseas Operations," 16 Dec 1940, in TSAGD, Microfilm Unit, 1941 Correspondence, Reel 13, Item 11.

2. WD GO No. 9, 22 Jun 1927.

3. Secret Cable No. 87 from Maj Gen George H. Brett, to the War Department, 25 Sep 1941, in Supporting Documents of "History of the Eighth Air Force Service Command."

4. *Ibid.*

5. GO No. 64, HG Panama Canal Department, Quarry Heights, Canal Zone, 14 Dec 1940.

6. Interview with Maj Gen E. E. Adler, 16 Feb 1945. Also "Brief History of the 1st Air Depot Group, 1 Jan 1941 to 7 Dec 1941," and "Brief History of Hq and Hq Squadron, 1st Air Depot Group, 1 Jan 1941 to 7 Dec 1941." For further information on the background of this situation, see *U.S. Naval Logistics in the Second World War*, by Duncan S. Ballantine, pp. 62–63.

7. Memo for the Chief of Air Corps by Brig Gen George H. Brett, Acting Chief of Air Corps, "Depot Supply and Repair Squadrons for Overseas Operations," 16 Dec 1940, in TSAGD File, Microfilm Unit, 1941 Correspondence, Reel 13, Item 11.

8. *Ibid.*

9. *AAF, The Official Guide to the Army Air Forces*, p. 356.

10. Bldgs and Grounds Div, Washington, D.C. to TAG, "Supplemental Estimates F. Y. 1941," 26 Dec 1940, TSCON File 600.121 (General).

11. For further information on construction, see ATSC Historical Monograph, "Acquisition of Facilities for the Air Service Command."

12. Chief, Civilian Personnel Div, to Chief, Bldgs & Grounds Div, "Request for Inclusion of Eleven Blitzkrieg-type Hangars in Budget Estimate," 23 Apr 1941. In TSAGD File, Microfilm Unit, 1941 Correspondence, Reel 13, Item 11.

13. TAG to CO's, McClellan Field, California, Duncan Field, Texas, and Patterson Field, Ohio, "Air Depot Groups," 26 Mar 1941, in *ibid.*

14. *Ibid.*

15. Lt Col A. G. Liggett, Sacramento Air Depot, to Col Henry J. F. Miller, Provisional Air Corps, Maintenance Command, 10 Apr 1941, in *ibid.*

16. *Ibid.*

17. 1st Ind (CO, SAD, to Chief, Provisional Air Corps Maintenance Command, 11 Apr 1941), Lt Col G. C. Nutt, Provisional Air Corps Maintenance Command, 21 Apr 1941, in *ibid.*

18. Lt Col F. S. Borum, Chief, Field Service Section, to Fairfield Air Depot, 18 Apr 1941, copies also sent to San Antonio and Sacramento, in *ibid.*

19. "History of San Antonio Air Depot from Inception to 1 Feb 1943," p. 89.

20. "Organizational History of the Fourth Air Depot Group from 1 Jan to 31 Dec 1942."

21. TWX FS 503, 7 Apr 1941, in TSAGD, Microfilm Unit, 1941 Correspondence, Reel 13, Item 11.

22. *AAF, The Official Guide to the Army Air Forces,* p. 357.

23. Col Henry J. F. Miller, to Chief, Materiel Division, Washington, D.C., "Assignment of Enlisted Personnel to Air Depot Groups," 23 Apr 1941, in TSAGD, Microfilm Unit, 1941 Correspondence, Reel 13, Item 11.

24. *Ibid.*

25. 1st Incl: Brig Gen H. J. F. Miller, Chief, Maintenance Command, to Chief, Air Corps, through Chief, Materiel Division, "Transfer of 5th Air Depot Group," 20 Oct 1941, in *ibid.*

26. Col Henry J. F. Miller to Chief, Materiel Division, Washington, D.C., "Assignment of Enlisted Personnel to Air Corps Depot Groups," 23 Apr 1941, in *ibid.*

27. Col Henry J. F. Miller to Chief, Materiel Division, Washington, D.C., "Transfer of Enlisted Personnel to Mobile Air Depot Groups," 28 Apr 1941, in *ibid.*

28. TWX, Brig Gen O. P. Echols to Chief, FSS, 22 Jan 1941, in TSAGD File 323.61 (Organization—Air Service Command).

29. Col Henry J. F. Miller, Chief, Maintenance Command, to Maintenance Section, Materiel Division, OCAC, "Transfer of Enlisted Personnel to the

Mobile Air Depot Groups," 26 Jun 1941, in TSAGD, Microfilm Unit, 1941 Correspondence, Reel 13, Item 11.

30. This list had been forwarded to Col Miller by Lt Col Clements McCullen, San Antonio Air Depot, on 24 May 1941, in *ibid.*

31. Maj Gen George H. Brett, Chief of Air Corps, to the CG's, South East, Gulf Coast, and West Coast ACTC's, the ACTTC, and all stations under the direct control of the CAC, "Activities and Operations of the Air Corps Main. Command." 28 Aug 1941, in TSHIS—2 files.

32. Brig Gen Henry J. F. Miller, Chief, Maintenance Command, to Col Clements McMullen, CO, San Antonio Air Depot, "My dear Mac," 8 Sep 1941, in TSAGD, Microfilm Unit, 1941 Correspondence, Reel 13, Item 11.

33. Col Henry J. F. Miller, CO, PACMC, to The Adjutant General of the Army, through Chief, Materiel Division, "Assignment of Specialized, Qualified Selectees to the Air Depot Groups," 29 Apr 1941, in *ibid.*

34. TAG 324 (4–21–41) E-A, "Procurement of Selective Service Men," 24 Apr 1941.

35. Maj J. C. Gordon to CO, McClellan Field, "Selectees Occupational Specialist Requirements for Air Depot Groups, Air Corps," 4 Aug 1941, in TSAGD File, Microfilm Unit, 1941 Correspondence, Reel 13, Item 11.

36. Maj C. E. Thomas, Jr., to CO, 3rd Air Depot Group, through CO, San Antonio Air Depot, "The Assignment of Specialized, Qualified Selectees to Air Depot Groups," 5 May 1941, in *ibid.*

37. 1st Lt Martin A. Bateman, Adjutant, San Antonio Air Depot, to The Adjutant General, War Department, Washington, D.C., "Special Report on Organization and Activation of Units," 1 May 1941, in *ibid.*

38. "Organizational History of the Fourth Air Depot Group from 1 Jan to 31 Dec 1942."

39. SO No. 1, Hq WAD, 2 Jan 1942, and SO No. 4, Hq WAD, 6 Jan 1942.

40. SO No. 3, Hq WAD, 5 Jan 1942, SO No. 12, Hq WAD, 19 Jan 1942, and SO No. 14, Hq WAD, 21 Jan 1942.

41. Col Henry J. F. Miller, Chief, Maintenance Command, to CO, San Antonio Air Depot, "Air Depot Groups," 26 Jun 1941, in Microfilm Unit, 1941 Correspondence, Reel 13, Item 11.

42. Lt Col F. S. Borum, Chief, FSS, to CO, Fairfield Air Depot, 18 Apr 1941, in *ibid.* Copies were also sent to San Antonio and Sacramento.

43. *Ibid.*

44. Memo for the Chief of Air Corps by Brig Gen George H. Brett, Acting Chief of Air Corps, "Depot Supply and Repair Squadrons for Overseas Operations," 16 Dec 1940, in TSAGD, Microfilm Unit, 1941 Correspondence, Reel 13, Item 11.

45. "Organizational History of the Fourth Air Depot Group from 1 Jan to 31 Dec 1942."

46. *Ibid.* See also the "History of the Eighth Air Force Service Command."

47. Memo for Chief, FSS, by John M. des Islets (for Lt Col E. E. Adler),

"Facilities for Overhaul of Engines and Engine Accessories by Air Depot Groups," 24 Sep 1941, in TSAGD, Microfilm Unit, 1941 Correspondence, Reel 13, Item 11.

48. Memo for Chief, Maintenance Command, by Lt Col Joseph T. Morris, "Overhaul of Engines and Engine Accessories by Air Depot Group," 30 Oct 1941, in *ibid.*

49. *Ibid.*

50. T/O 1–917, 1 Jul 1943.

51. Col William O. Butler, Assistant Chief, ASC, to Chief of the Air Corps, "Augmentation of Air Depot Groups," 8 Dec 1941, in TSAGD, Microfilm Unit, 1941 Correspondence, Reel 13, Item 11.

52. Lt Col Joseph H. Hicks, Chief, FSS, to Supervisors Concerned, "Engine Overhaul Equipment for Air Depot Groups," 19 Jan 1942, in TSPER 322 (Air Depot Groups—General).

53. *Ibid.*

54. Interview, Mr. Paul Angle with Brig Gen E. E. Adler, 10 March 1944. For further information on the maneuvers, see *The Organization of Ground Combat Troops* by Kent Roberts Greenfield, Robert R. Palmer, and Bell I. Wiley, pp. 109–110.

55. Maintenance Command Headquarters, Lt Col E. E. Adler, Chief, Plans Division, Memorandum to All Concerned, "Operating Instructions Relative to Air Force Combat Command Maneuvers, September–November 1941," 14 Aug 1941, in TSHIS-2 files.

56. *Ibid.*

57. *Ibid.* The "Memorandum to All Concerned" was an SOP applicable to civilian and military personnel alike.

58. Memo for Chief, Maintenance Command, by Maj Gen George H. Brett, Chief of the Air Corps, "Participation in September–November Maneuvers, AFCC," 15 Aug 1941, in TSHIS-2 files.

59. Interview with Maj A. E. Freathy, TSMCO, on 17 Jan 1945.

60. *Ibid.*

61. *Ibid.* Major Freathy was the Warrant Officer referred to in the text.

62. Memo for Chief, FSS, by Lt Col Joseph T. Morris, "Report on Maneuvers," 14 Oct 1941, in TSMCO file 323.3 (Air Depot Group, Fourth).

63. Memo for the Chief of the Air Corps, by Lt Col Barney M. Giles, "Special Insp. of AAF Units Participating in Manuevers, September 14–30, 1941," 3 Oct 1941, in TSHIS-2 files.

64. MC-T-9861, 29 Sep 1941, in TSMCO file 323.3 (Air Depot Group, Fourth).

65. T/O&E's 1–852 and 1–857, 1 Jul 1943.

66. Memo for the Chief, Maintenance Command, by Maj Max H. Warren, "Report on the Activities of the 4th Air Depot Group in the Louisiana Maneuvers," 27 Oct 1941, in TSAGD, Microfilm Unit, 1941 Correspondence, Reel 13, Item 11.

67. T/O's 1–857 and 1–858, 1 Jul 1943.

68. Memo to the Chief of the Air Corps, by Lt Col Barney M. Giles, "Special Insp. of AAF Units Participating in Maneuvers, September 14–30, 1941," 3 Oct 1941, in TSHIS-2 files.

69. CG, Air Force Combat Command, to CG, Third Air Force, "Maneuvers," 29 Aug 1941 (ACC 354.2, 8–29–41) in TSHIS-2 files.

70. Capt H. M. Price, Assistant Adjutant General, Hq, Air Force Combat Command, to Chief, Maintenance Command, "First Army—IV Corps Maneuvers," 2 Sep 1941 (ACC 354.2, 9–2–41) in TSHIS-2 files.

71. CG, Air Force Combat Command, to CG, Third Air Force, "Maneuvers," 29 Aug 1941 (ACC 354.2, 8–29–41) in *ibid.*

72. "History of Herbert Smart Airport from Activation to 1 Jan 1944."

73. Interview with Maj A. E. Freathy, TSMCO, on 17 Jan 1945.

74. Statement by Maj Max Warren in transcript of telephone conversation between Maj Warren (Jackson, Mississippi) and Maj des Islets and Lt Boyles (Hq ASC) on 25 Sep 1941. In TSMCO 323.3 (Air Depot Group, Fourth).

75. ASC-T-51, 17 Nov 1941, in TSAGD, Microfilm Unit, 1941 Correspondence, Reel 13, Item 11.

76. Statement of the problems by Maj Warren (Jackson, Mississippi) and Maj des Islets and Lt Boyles (Hq ASC) on 25 Sep 1941. In TSMCO 323.3 (Air Depot Group, Fourth).

77. *Ibid.*

78. *Ibid.*

79. Interview with Maj A. E. Freathy, TSMCO, on 17 Jan 1945.

80. "History of Herbert Smart Airport from Activation to 1 Jan 1944."

81. Interview with Maj A. E. Freathy and transcript of telephone conversation between Maj Warren (Jackson, Mississippi) and Maj des Islets and Lt Boyles (Hq ASC) on 25 Sep 1941 in TSMCO 323.3 (Air Depot Groups, Fourth).

82. "Organizational History of the Fourth Air Depot Group from 1 Jan to 31 Dec 1942."

83. 1st Ind, Maj Edward P. Curtis, Secretary of the Air Staff, to Chief of the Air Corps, "Transfer of the 5th Air Depot Group," 22 Oct 1941, in TSAGD, Microfilm Unit, 1941 Correspondence, Reel 13, Item 11.

84. Interview with Col J. D. Howe on 28 Apr 1945.

85. "History of Warner Robins Air Service Command to 1 Jan 1944," Vol. II, Section on Herbert Smart Airport.

86. "History of the San Antonio Air Service Command from Inception to 1 Feb 1943," p. 61.

87. Transcript of telephone conversation between Col L. P. Whitten (OCAC) and Lt Col J. T. Morris (Wright Field) 8 Dec 1941, in TSMCO file 323.3 (Air Depot Groups, Fourth).

88. *Ibid.*

89. *Ibid.*

90. Quoted in the "Organizational History of the Fourth Air Depot Group, 1 Jan to 31 Dec 1942."

91. TWX SCL-711, 8 Dec 1941 (received at Division Headquarters, Wright Field, at 8:41 PM) in TSMCO file 323.3 (Air Depot Groups, Fourth).

92. Same as No. 82, above. See also *The Army Air Forces in World War II*, Vol. I, pp. 422–3.

93. Col L. T. Miller, Assistant Chief, ATC, to CG, ASC, 19 Mar 1942, in TSAGD 322 (Organization—Groups).

94. "History of San Antonio Air Depot from Inception to 1 Feb 1943," p. 89.

## CHAPTER 3

1. Teletypes: ASC-T 2156, 2314, 2315, 2316 from the Air Service Command to the CO's of San Antonio, Sacramento, Middletown, Mobile, and Ogden Air Depots, 8 Dec 1941.

2. Memo, Office of the Chief, ASC, to Chief of Staff, "Housing 40-Depot Groups Program," 29 Dec 1941, in TSCON 600.121 (Depot Groups).

3. TAG 320.2 (12–21–41) MR-M-AAF, "Constitution and Activation of certain Air Corps Units," 30 Dec 1941.

4. TAG 320.2 (12–27–41) MR-M-AAF, "Constitution and Activation of Air Depot Groups," 5 Jan 1942.

5. TAG 320.2 (12–27–41) MR-M-AAF, "Constitution and Activation of Air Depot Groups," 31 Jan 1942.

6. "Development of the Air Force Logistical System," p. 7. For further information on the expansion of the Air Corps, see *The Army Air Forces in World War II*, Vol. I, pp. 104–116, 245–267.

7. ASC Memo. No. 50–2, "Master Training Program for Air Depot Groups," 17 Feb 1942.

8. Memo, re Col Aubrey Moore, 22 Dec 1941, in TSPER 322 (Air Depot Groups—General).

9. For further information see ATSC Historical Monograph "Requirements."

10. Col E. H. White, Budget Officer to CG, Materiel Command, 31 Jul 1942, "Procurement of Additional Airplane Equipment," in TSCBO 400.35 (1942).

11. Memo by Lt Col W. B. Bigson, Chief, Requirements Branch, Supply Division, "Policy Changes," 28 Apr 1944, in TSHIS—2 files. See also Six Month Troop Forecasts, or Commitment Charts, issued by AC/AS, OC&R.

12. Col W. O. Butler, Assistant Chief, Air Service Command, to Chief, Air Service Command, 21 Feb 1942, in TSAGD File 322 (Organization—Groups). See also Victory Program Equipment Cost Sheet, 12 Feb 1942, issued by Lt Col Joseph H. Hicks, Jr., Chief, FSS, in TSAGD 381 (Victory Program).

13. See section on "Sources of Enlisted Personnel."

14. For further information see "History of the Acquisition of Facilities for the Air Service Command," pp. 32–36 and 98–104.

15. R&R, Engineer Section to Chief, FSS, "Housing for Air Depot Groups," 16 Feb 1942, in TSCON 600.121 (Depot Groups).

16. 5th Ind, Hq AAF, to Chief of Engineers, 23 Feb 1942, "Mobile Depot Groups," in TSCON 600.121 (General).

17. R&R, Director of Base Services, Buildings and Grounds, to ASC, "Housing for Air Depot Groups Personnel," 11 Mar 1942, in *ibid.*

18. 1st Ind, Inter-office Memo, Col Ralph Nemo to CG, ASC, "Housing for Air Depot Groups Personnel," 23 Mar 1942, in TSAGD 322 (Organization—Groups).

19. ASC drawings 100s-2, 100s-3, and D-1006–1. (Specifications were later revised 1 May 1942 and 12 Jun 1942.)

20. R&R, Engineer Section to Chief of Staff, "Construction Cost for Camps," 31 Mar 1942, in TSCON 600.121 (Depot Groups—Oct 1941 to Mar 1942). See Supporting Documents for statement of the cost of Air Depot Group facilities.

21. R&R, Engineer Section to Chief, FSS, "Housing for Air Depot Groups," 2 Feb 1942, in TSCON 600.121 (Depot Groups—Oct 1941 to Mar 1942).

22. Contracts for construction of prefabricated buildings were let with Alladin Co., Bay City, Michigan, Texas Prefabricating House & Tent Co., Dallas, Texas, T. C. King Co., Anniston, Alabama, and several other concerns. Engineer Section to Control Section, "Authority for Purchase," 9 May 1942, in TSCON 600.121 (Depot Groups).

23. R&R, A-4-F to Engineer Section, "Buildings for the 13th Air Depot Group," 14 Aug 1942, in TSCON 600.121 (Depot Groups, Jul–Sep 1942).

24. Col L. T. Miller, Chief of Staff, to CG, ASC, 2 Jan 1942, in TSPER 322 (Air Depot Groups—General).

25. See Supporting Documents for list, and copies, of training manuals issued by Headquarters, ASC.

26. For the organization of an Air Depot Group in May 1942, see "Preliminary Instruction Manual, Organization and Functions, Hq and Hq Squadron, Air Depot Group."

27. See "Index to AAF Tables of Organization and Equipment, Tables of Organization, Tables of Distribution, and Manning Tables" in Supporting Documents.

28. Memo, 17 Feb 1942, in TSAGD 353 (Training).

29. TAG 320.2 (12–18–41) MR-M-AAF, "Constitution and Activation of Certain Air Corps Units," 27 Dec 1941, and AAF Reg. No. 65–6, 7 Feb 1942.

30. R&R, Military Personnel Section, to Col F. M. Zeigler, "Report of Enlisted Personnel, ASC," 23 Apr 1942, in TSCON 600.121 (Depot Groups).

31. Maj C. E. Thomas, Jr., to CO, 3rd ADG, thru: CO, San Antonio Air Depot, "The Assignment of Specialized, Qualified Selectees to Air Depot

Groups," 5 May 1941, in TSAGD, Microfilm Unit, 1941 Correspondence, Reel 13, Item 11.

32. TAG to Chief, AAF, CG, AF Combat Command, *et al*, 10 Dec 1941, in TSPER 327.02 (Selective Service).

33. AAF Reg. No. 15–127 and 15–128, both dated 23 Feb 1942.

34. 3rd Ind, Col J. M. Bevans, Director of Personnel (by Maj T. M. Belshe), to CG, Services of Supply, 24 Jun 1942, in TSAGD 322 (Organization—Air Depot Groups).

35. AAF Reg. No. 15–127, 15 Sep 1942.

36. *Ibid.*

37. Memo, "Changes of Policy and Procedure Pertaining to the Assignment of Enlisted Men Since Establishment of the Maintenance Command," prepared by Lt Col Havens, ASCPM-2, 1 Aug 1944, in TSHIS-2 files.

38. Report of Investigation of the Air Service Command; Annual General Inspection, 1 Sep 1943, in TSAIR files.

39. Col L. T. Miller to CG, ASC, "Report on the 24th and 25th Air Depot Groups," 24 Apr 1942, in TSAGD 322 (Air Depot Groups).

40. Col L. T. Miller, Assistant Chief, ASC, to CG, ASC, 19 Mar 1942, in *ibid.*

41. "History of the Officers Branch, Military Personnel Section," a memorandum in TSHIS-2 files.

42. AR 140–23, 5 Jul 1940.

43. AR 140–39, 24 Jul 1933.

44. Memorandum, 9 Dec 1941, in TSPER 210.105 (Candidates for Commissions).

45. Interview with Capt Nancy P. Julia, 25 Apr 1945.

46. Col William W. Dick, Air Adjutant General, to Appointment and Procurement Division, Air Staff, A-1, "Establishment of Liaison," 20 Apr 1942, in TSPER 210.105 (Appointment of Key Civilians in the AUS).

47. TAG 210.1 (5–6–42) RB-SPGA, 8 May 1942.

48. TAG 210.31 (8–4–42) RB-A, 8 Aug 1942.

49. TWX ASC-T-579-A-IM, 25 Sep 1942, in TSPER 210.105 (Appointment of Key Civilians in the AUS).

50. Capt Weldon B. Gibson to Mr. John W. Chandley, Evansville, Indiana, 30 Sep 1942, in TSPER 210.105 (Appointment of Key Civilians in the AUS).

51. Maj Leo C. Wilson, Acting Assistant Adjutant General, to The Adjutant General, Washington, D.C., "Officer Requirements for Arms and Services for Air Depot Groups," 4 Apr 1942, in TSAGD 322 (Air Depot Groups).

52. AAF Memo No. 10–1, 21 May 1942, which provided that officers should sign their names "1st Lt., A.A.F." was rescinded by AAF Memo No. 10–1, 25 Jun 1942. Thereafter branch of the service was written thus: "1st Lt., A. C."

53. AAF Reg. No. 15–127, 15 Sep 1942.

54. Interview with Maj W. R. Miller, 25 Apr 1945.

55. AAF Memo No. 35–8, 4 Nov 1942.
56. History of OCS, *Class Book, Class 44D.*
57. Chief, Field Services, to CO, 2nd ASAC, 11 Aug 1942, in P&BS Officers Branch Policy Book.
58. "History of Warner Robins Air Service Command to 1 Jan 1944," Vol. II, Section on "The Air Service Command School."
59. ASC Memo 50–2, 17 Feb 1942, "Master Training Program for Air Depot Groups."
60. 3rd Ind (Col Harold A. McGinnis, CO, 8th AFSC to CG, ASC), Col Clements McMullen, Chief Overseas Division, to 8th AFSC, "Equipment for Warton Air Depot, United Kingdom," 26 May 1942, in TSAGD 322 (Air Depot Groups).
61. *Ibid.*
62. Col T. J. Hanley, Jr., "Memorandum for Maj Gen H. J. F. Miller," 19 Mar 1942, in TSAGD 322 (Organization—Groups).
63. For a definition of a depot see AAF Reg. No. 85–5.
64. Lt Col Ralph Nemo to CG, ASC, "Memorandum from Colonel T. J. Hanley, Jr.," 28 Mar 1942, in TSAGD 322 (Organization—Groups).
65. Lt Col Joseph H. Hicks, Chief, FSS, to Supervisors Concerned, "Engine Overhaul Equipment for Air Depot Groups," 19 Jan 1942, in TSPER files, 322 (Air Depot Groups—General).
66. *Ibid.*
67. Col T. J. Hanley, Jr., "Memorandum for Maj Gen H. J. F. Miller," 19 Mar 1942, in TSAGD 322 (Organization—Groups).
68. Same as No. 64, above.
69. For further information see ATSC Historical Monograph "Supply of AAF Organizational Equipment to the Army Air Forces."
70. ASC Memo No. 65–3, 11 Jul 1942.
71. AAF Reg. No. 15–106 and AAF Reg. No. 65–12.
72. *Ibid.*
73. WD Cir. No. 105, 1942.
74. ASC Memo No. 65–3, 11 Jul 1942.
75. AAF Reg. No. 65–12, 11 May 1942.
76. For further information see ATSC Historical Monograph "AAF Supply of Overseas Air Forces."

## CHAPTER 4

1. See ASC Memo 50–2, "Master Training Program for Air Depot Groups," 17 Feb 1942.
2. "History of Sacramento Air Service Command," p. 841.
3. Interview with Mr. Norman Hamilton, 28 Apr 1945.
4. ASC No. 4, 9 Dec 1941, effective Dec 1941.
5. See Section on "Technical Training."
6. See "Annual Report of the Instructional Property Branch," 3 Nov 1944.

7. See Speech of Brig Gen Charles E. Thomas, Jr. at ASC Conference on 4, 5, and 6 Oct 1943 in "The History of Warner Robins Air Service Command to 1 Jan 1944," Vol. II.

8. Interview with Maj Gen E. E. Adler on 22 May 1945. For details on the number of untrained personnel that were received see section on "Technical Training."

9. *Ibid.*

10. Interview with Maj J. B. League, 22 Feb 1945.

11. R&R, Col James D. Givens, G-3, to G-1, 18 Jun 1942, in TSHIS-2 files.

12. *Ibid.*

13. R&R, Chief of Staff to G-3, 30 Dec 1941, in TSPER 322 (Air Depot Groups—General).

14. ASC Memo 50–2, "Master Training Program for Air Depot Groups," 17 Feb 1942.

15. *Ibid.*

16. See correspondence in re: 36th Air Depot Group, in TSPER 322 (Air Depot Groups—General).

17. Par 3, Memo, Office of the Assistant Chief, Air Service Command, 16 Mar 1942, "Squadron Training Progress Report," in TSHIS-2 files.

18. Interview with Mr. Norman Hamilton, 28 Apr 1945.

19. ASC Memo No. 50–2, "Master Training Program for Air Depot Groups," 17 Feb 1942.

20. See ASC GO No. 29, 8 Apr 1942, and subsequent GO's.

21. ASC GO No. 42, 12 May 1942.

22. ASC GO No. 46, 25 May 1942.

23. ASC GO No. 55, 23 Jun 1942.

24. ASC GO No. 56, 24 Jun 1942.

25. ASC GO No. 66, 11 Jul 1942.

26. For further information on construction at these twenty-two locations, see section on "Acquisition of Facilities."

27. Col Leo H. Dawson to CG, ASC, Attn: Brig Gen McMullen, "Air Depot Group Training," 10 Sep 1942, in TSPER 322 (Air Depot Groups—General).

28. Col James D. Givens to Col Leo Dawson, "Dear Leo," 14 Sep 1942, in *ibid.*

29. "Development of Air Force Logistical System."

30. ASC Memo 50–2, "Master Training Program for Air Depot Groups," 17 Feb 1942.

31. This discussion is based on the 20th Station Complement at Herbert Smart Airport. For further information, see the "History of Warner Robins Air Service Command to 1 Jan 1944."

32. Interview with Col Russel Scott, 19 Jan 1944.

33. See "Organizational History of the 26th Air Depot Group from Date of Activation to 1 Sep 1943," and other unit histories.

34. *Ibid.*

35. For further information see ATSC Historical Monograph "History of the Procurement and Training of Civilian Personnel, Air Service Command," Parts I & II.

36. R&R, Cmt. 4, G-3 to G-1, 19 Feb 1942, in TSPER 322 (Air Depot Groups—General).

37. A detailed list of the Service Schools, Contract Schools, and Factory Schools to which ASC personnel were sent, complete with the number of trainees at each course in 1942, 1943, and 1944 is included in the Supporting Documents.

38. R&R, Cmt. 4, G-3 to G-1, 19 Feb 1942, in TSPER 322 (Air Depot Groups—General).

39. Contracts were made by the Materiel Command, not only for the Air Service Command, but also for the other commands of the Army Air Forces.

40. See ASC GO No. 27, 26 Mar 1942.

41. Maj Gen W. H. Frank, CG, ASC, to CG, AAF, 8 May 1943, in AAG 161 (Contract Schools). For further information see AAF Historical Study No. 26.

42. Interview with Lt Col J. L. Reass, 2 Mar 1945.

43. See ASC GO No. 27, 26 Mar 1942.

44. TWX ASC-T-936 G-3, 8 May 1942, in TSHIS-2 files.

45. *Ibid.*

46. *Ibid.*

47. "History of the Sacramento Air Service Command," p. 862.

48. Col William W. Dick to CO's of the AAF in the Continental U.S., 25 Jun 1942, in TSPCP Civilian Training Program II.

49. Maj Gen W. H. Frank, CG, ASC, to CG, AAF, 8 May 1943, in AAG 161 (Contract Schools). For further information see AAF Historical Study No. 26.

50. ASC Memo No. 50–2, "Master Training Program for Air Depot Groups," 17 Feb. 1942.

51. *Ibid.*

52. For further information see Restricted Manual, "Organization and Training of an Air Depot Group, Patterson Field, Fairfield, Ohio," 15 Sep 1942, revised to Dec 1st, and similar manuals in Supporting Documents.

53. TWX ASC-T-367A3, 25 Feb 1942, in TSHIS-2 files.

54. *Ibid.*

55. R&R from Ordnance Section to G-3, ASC, 8 Jun 1942, in TSPER 322 (Air Depot Groups—General).

56. *Ibid.*

57. Interview with Col Russel Scott on 20 Jan 1944.

58. For further information see "The History of Warner Robins Air Service Command to 1 Jan 1944," Part II.

59. Memo for General Stratemeyer by Lt Gen Joseph T. McNarney,

Deputy Chief of Staff, "Training Service Units," 31 Dec 1942, in WDCSA 320.2 (12/13/42) in TSHIS-2 files.

## CHAPTER 5

1. See "The History of Rockwell Field" by Maj H. H. Arnold, Air Service, U.S. Army, 1923.
2. See Section on "The Air Corps and the GHQ Air Force."
3. Condensed from "History of the 3rd Service Group, Activation Date to Overseas Date."
4. Lecture Manuscript, "Theatre Administration, Air Force Presentation at a War Department Conference, 8 Feb 1944," by Brig Gen L. P. Whitten, Chief, Air Services Division AC/AS, MM&D, p. 32.
5. *Ibid.*
6. HASC 302C, 14 Apr 1942, in TSHIS-2 files.
7. "The Air Force in Theatres of Operations: Organizations and Functions," an AAF Manual dated May 1943.
8. Teletype ASCT-943G3, 16 Apr 1942, in TSHIS-2 files.
9. R&R, AFASC to AFRBS (1st Provisional Service Group, Fort Dix, N.J.).
10. Memo for CG, ASC, by Col L. P. Whitten, Director of Base Services, "Service Test of Air Depot Group," 18 Apr 1942, in TSAGD 353 (Training).
11. For a complete list of aircraft serviced by this group, see Lt Col W. E. Steele, Chief, Supply Division to CO, 3rd ASAC, "Report on First Service Group, Fort Dix, N.J.," 12 Jun 1942, in TSPER (1st Provisional Service Group, Fort Dix, N.J.).
12. Memo for the Director of Base Services, through CG, ASC, by Lt Col P. E. Ruestow, Lt Col H. P. Dellinger, and 2nd Lt R. R. Scott, 21 Apr 1942, in TSAGD 322 (Air Depot Groups).
13. *Ibid.*
14. *Ibid.*
15. "Memorandum for All Concerned" by Lt Col H. Paul Dellinger, "Conference on Experimental Air Depot Group," 25 Apr 1942, in TSAGD 322 (Air Depot Groups).
16. *Ibid.*
17. In Supporting Documents to the "History of the Eighth Air Force Service Command."
18. *Ibid.*
19. Interview with Lt Col C. R. Hitchcock, 8 Mar 1945.
20. *Ibid.*
21. Lt Col H. Paul Dellinger to Chief of Staff, ASC, "Preliminary Report on Functioning of Provisional Service Group," 16 May 1942, in TSPER (1st Provisional Service Group, Fort Dix, N.J.).
22. *Ibid.*

23. Lt Col H. Paul Dellinger to Chief of Staff, Field Services, "Report on Functioning of Provisional Service Group," 16 Jun 1942, in *ibid.*
24. *Ibid.*
25. *Ibid.*
26. *Ibid.*
27. Same as No. 21, above.
28. *Ibid.*
29. Maj Gen Henry J. F. Miller, to CG, AAF, "Revision of Maintenance and Supply Plans," 19 May 1942, in *ibid.*
30. Lt Col H. Paul Dellinger to Chief of Staff, Field Services, "Report on Functioning of Provisional Service Group," 16 Jun 1942, in *ibid.*
31. For list of training manuals available to this group for instructional purposes, see List of Instructional Manuals, 10 Jul 1942, in *ibid.*
32. Lt Col W. E. Steele, Chief, Supply Division, to CO, 3rd ASAC, "Report on First Service Group, Fort Dix, N.J.," 12 Jun 1942, in *ibid.*
33. *Ibid.*
34. Col William W. Dick, AAG, to all Post and Station Commanders and Higher Headquarters, AAF, in the continental United States, "Reorganization of Service Units of the Army Air Forces," 5 May 1942, in TSAGD 322 (Air Depot Groups).
35. *Ibid.*
36. *Ibid.*
37. Cochran Field Sub-depot TWX-1670, 1 Jun 1942, in TSHIS-2 files.
38. *Ibid.*
39. Teletype ASCT-10000A2, 27 May 1942, in TSHIS-2 files.
40. Teletype ASCT-1215G3, 10 Jun 1942, in TSHIS-2 files.
41. *Ibid.*
42. TWX ASCT-785A3, 26 Jun 1942, in *ibid.*
43. 3rd ASAC-T-6-X-244, 13 Jun 1942, in *ibid.*
44. TAG 320.2 (6–10–42) MR-M-AF, 13 Jun 1942.
45. TAG 320.2 (7–10–42) MR-M-AF, 14 Jul 1942.
46. See Unit Historical Files, TSCSU-2A.
47. TAG 320.2 (6–14–42), MR-AF-TS-M, 13 Jun 1942.
48. Col William W. Dick, AAG, to CG's, all Air Forces in continental U.S. and CG's, all Army Air Force Commands, "Responsibility for Service Units," 22 Jun 1942, in TSAGD 322 (Organization—AAF Units).
49. For further disposition of these units made by the Air Service Command, see ASC GO No. 58, 27 Jun 1942; ASC GO No. 62, 5 Jul 1942; ASC GO No. 87, 15 Aug 1942; ASC GO No. 90, 19 Aug 1942; and other appropriate directives.
50. *Ibid.*
51. Col William W. Dick, AAG, to CG's in continental U.S. and CG's, all Army Air Force Commands, "Responsibility for Service Units," 22 Jun 1942, in TSAGD 322 (Organization—AAF Units).

52. Lecture Manuscript, "Theatre Administration, Air Force Presentation at a War Department Conference, 8 Feb 1944," by Brig Gen L. P. Whitten, Chief, Air Services Division, AC/AS, MM&D.

53. *Ibid.*

CHAPTER 6

1. Presentation by Col Enter, ASC, in "Panel Discussion by the Chief of Staff and Special Staff Representatives of the Air Service Command at AAFSAT," 27 Sep 1943, hereafter cited as "Panel Discussion . . . at AAFSAT."

2. See "The Unit Gas Officer and Gas Non-Commissioned Officer," in Appendix C, Supporting Documents. This manual was prepared by the CWS Section, Hq ASC, and later became Air Forces Manual No. 50.

3. Presentation by Col Enter, ASC, in "Panel Discussion . . . at AAF-SAT."

4. Col William W. Dick, AAG, to CG's, all Air Forces in continental U.S. and CG's, all Army Air Force Commands, "Responsibility for Service Units," 22 Jun 1942, in TSAGD 322 (Organization—AAF Units).

5. GO No. 82 and 88, Hq ASC, Jul 1942.

6. Memo prepared by Maj A. S. DeRossett, TSCWS, "Chemical Warfare Units with the ATSC."

7. Presentation by Col Enter, ASC, in "Panel Discussion . . . at AAF-SAT."

8. *Ibid.*

9. *Ibid.*

10. Memo prepared by Maj A. S. DeRossett, TSCWS, "Chemical Warfare Units with the ATSC."

11. *Ibid.*

12. *Ibid.*

13. *Ibid.*

14. Complete statistical information on CWS units trained by the Air Service Command is available in TSHIS-2 files.

15. Presentation by Col Cotter, ASC, in "Panel Discussion . . . at AAFSAT."

16. *Ibid.*

17. Brig Gen S. C. Godfrey, Air Engineer, to CG, AAF, "Aviation Engineer Program," 25 Oct 1943, in AFDAE-1 files.

18. *Ibid.*

19. Presentation by Col Cotter, ASC, in "Panel Discussion . . . at AAFSAT."

20. "Organizational Development, Northeast Air District and First Air Force, 19 Nov 1940 to 31 Dec 1943," p. 22.

21. Brig Gen S. C. Godfrey, Air Engineer, Memo for the Chief of Air Staff,

"Assignment and Basic Training of Aviation Engineers," 20 May 1943, in AFDAE-1 files.

22. *Ibid.*

23. *Ibid.*

24. Report of the Historical Division, Hq AAF, to WDGS, G-2, on Training in the AAF, chapter on "Training of Service and Maintenance Units."

25. Brig Gen S. C. Godfrey, Air Engineer, to CG, AAF, "Aviation Engineer Program," 25 Oct 1943, in AFDAE-1 files.

26. *Ibid.*

27. "Arms and Services Training in the First Air Force," AG 322.03A (Corps of Engineers), in AAG Bulk # 66.

28. Brig Gen S. C. Godfrey, Air Engineer, to CG, AAF, "Aviation Engineer Program," 25 Oct 1943, in AFDAE-1 files.

29. See files on Camouflage Schools, in AFACT-4 files.

30. See "Camouflage Training Manual for the Air Service Command," in Appendix C, Supporting Documents.

31. Memo prepared by Miss E. Jones, TSBFO, "Finance Sections, Air Depot & Air Service Groups."

32. *Ibid.*

33. Presentation by Maj Leslie, ASC, in "Panel Discussion . . . at AAFSAT."

34. *Ibid.*

35. Same as No. 24, above.

36. Same as No. 31, above.

37. Col D. B. Schannep, Chief T&O, P&T Division, to CG, Warner Robins ASC, "Training Finance Personnel," 2 Jun 1943, in TSAGD 353 (Training).

38. *Ibid.*

39. Same as No. 31, above.

40. Presentation by Col White, ASC, in "Panel Discussion . . . at AAF-SAT."

41. This discussion is based on Section VIII, "Training," prepared by Maj Richard J. Brightwell, in "Medical History, Air Service Command, 1942–1945," forwarded to the Air Surgeon, copy in TSHIS-2 files. All information not otherwise acknowledged has been drawn from this report.

42. WD File AG 370.5 (7–12–43) OB-S-E-GN-AF-SPMOT-M, dated 1 Aug 1943, short title POM; and WD File AG 370.5 (8–6–43) OB-S-AF-M, dated I Aug 1943, short title Air POM.

43. Established by Annex No. 8, 11 Sep 1942, to ASC Reg. No. 50–2, 20 Aug 1942.

44. *Ibid.*

45. For further information see Medical Report in "The History of Warner Robins Air Service Command to 1 Jan 1944," Vol. II.

46. AAF Reg. No. 20–28, 15 Nov 1943.

47. See Medical Report in "The History of Warner Robins Air Service Command to 1 Jan 1944," Vol II.

48. *Ibid.*

49. P&T Division Cir. No. 50–140–3, 29 Dec 1943.

50. WD File AG 320.2, 17 Jul 1943, "Military Police Units for the Army Air Force."

51. WD File AG 320.2 (9–26–41) MR-M-A, 26 Sep 1941; WD File AG 320.2 (7–13–42) MR-M-WDGCT, 17 Jul 1942.

52. Maj T. M. Belshe for Col J. M. Bevans, Director of Personnel, to CG, AAF Foreign Service Concentration Command, Union Central Life Insurance Annex, "Guard Squadrons for Army Air Force Stations," 20 Aug 1942, in TSACD 353 (Training).

53. WD File AG 320.2 (6–10–42) MR-M-AF, 13 Jun 1942 and WD, File AG 221 (7–23–42) EA-AF-TS-M, 27 Jul 1942, cited in *ibid.*

54. Section on AAF Guard School, "The History of San Antonio Air Service Command 1 Feb 1943 to 1 July 1944," p. 81.

55. Report of Historical Division to WDGS, G-2, on Training in the AAF, Chapter on "Training of Service and Maintenance Units."

56. AFRIT to CG, AFTTC, 1 Aug 1942, in AAG 353.9E (Training—General).

57. See presentation of Col Robbins, ASC, in "Panel Discussion . . . at AAFSAT."

58. *Ibid.*

59. *Ibid.*

60. Col William W. Dick, AAG, to CG's, all AF's in continental U.S. and CG's, all Army Air Force Commands, "Responsibility for Service Units," 22 Jun 1942, in TSACD 322 (Organization—AAF Units).

61. "Personnel and Training Activities of the Ordnance Section from 1 June 1942 through 1 November 1943," prepared by Lt Col Joseph H. Bryson.

62. *Ibid.*

63. *Ibid.*

64. Complete statistical information on Ordnance Units trained by the Air Service Command is available in TSHIS-2 files.

65. Same as No. 55, above.

66. Inter-Office memo for AFRBS by Air Ordnance Officer, 1 Mar 1943, in AAG Bulk # 90.

67. Col D. B. Schannep, Chief, T&O Section, to CG, AAF, AC/AS, MM&D, Ordnance Officer, "Training of ASC Ordnance Units at ASF Installations," 12 Aug 1943, in TSACD 353 (Training).

68. See presentation by Col Malone, ASC, in "Panel Discussion . . . at AAFSAT."

69. *Ibid.*

70. *Ibid.*

71. Lecture at the Air Service Command School by the former Air Quartermaster, Twelfth Air Force.

72. Same as No. 60, above.

73. 1st Ind (Col E. T. M. Cavanaugh for Col H. R. W. Herwig, Air Quartermaster, "Elimination or Combination of Units Whose Functions are Duplicating or Overlapping," 24 Jul 1943) Col D. B. Schannep, Chief, T&O Section, to AC/AS, MM&D, Attn Air Quartermaster, 30 Jul 1943, in TSAGD 322 (Organization—AAF Units).

74. "Historical Report, Quartermaster Activities, ASC, ATSC," prepared by Lt Col R. Slough, 6 Mar 1945.

75. See remarks by Col Schannep in "Record of Conference Held at Hq, ASC, Patterson Field, Fairfield, Ohio, April 9th and 10th, 1943," in personal file of Maj J. M. McCampbell, TSPTR, "History."

76. Same as No. 74, above.

77. Maj William C. Hempfling to Chief, Unit Training Division, "Report of Visit to ASWAAF Activities, 10 Oct 1943," in Unit Training Division File, Hq AAF.

78. Same as No. 55, above.

79. Memo for Deputy for Services Training by Lt Col J. M. Faire, 14 Feb 1944, in TSHIS-2 files.

80. *Ibid.*

81. Same as No. 74, above.

82. 1st ASAC to Training Divisions, Quartermaster Organizations, 1 Sep 1942, in AAG 353.9C (Training Directives and Programs); "Arms and Services Training in the First Air Force," in AFIHI archives; Daily Activity Report, Air Quartermaster, 22 Nov 1944, in *ibid;* AC/AS, M&S, to CG, ATSC, 19 Dec 1944 in AAG 321H (Quartermaster).

83. For more detailed information see Chapter on Training in ATSC Historical Monograph, "Radio and Radar Supply, Maintenance, and Training by the Air Service Command."

84. See presentation of Maj Ludwig, ASC, in "Panel Discussion . . . at AAFSAT."

85. *Ibid.*

86. *Ibid.* Complete statistical information on Signal Corps units trained by the Air Service Command is available in TSHIS-2 files.

87. P&T Division Cir. No. 50–100–2, 22 Jan 1944.

88. Memo for Col L. H. Ross from Maj J. M. McCampbell, "VHF Personnel at Warner Robins Air Service Command," 24 Jun 1944, in personal file of Maj J. M. McCampbell (Memorandums—1944).

89. ASC GO No. 8, 17 Jan 1942. For further information see "Historical Data, Radio Section, 4000th AAF Base Unit (Command)."

90. See "Historical Report, VHF Engineering Agency, 1 February 1944."

91. See "History of the 1st Radar and VHF Installation and Maintenance Unit, 31 Aug 1943 to 31 May 1944."

92. The length of each course, the specific equipments covered, and the capacity of the school at Warner Robins are included in Appendix A, Exhibit

No. 12, ATSC Historical Monograph, "Radio and Radar Supply, Maintenance, and Training by the Air Service Command."

93. ASC Reg. No. 20–15, 1 Nov 1943.

94. R&R, Col L. S. Woods, Chief, QM Section, to ASCPM-2B, "Assignment of Enlisted Men," 15 Nov 1943, in ASCPM-2B 220.31.

95. See ATSC Historical Monograph, "Supply of AAF Organizational Equipment to the Army Air Forces," pp 40–56.

96. "Training and Morale of Arms and Services Units with the AAF," 15 Dec 1943, in TSAGD—Gen Frank file—353.01 (Training Program).

97. R&R, Maj W. R. Miller, Sr. to Lt Col Brown, through Lt Col Havens, "Assignment of Enlisted Personnel," 27 Aug 1943, ASCPM-2B 220.31.

CHAPTER 7

1. ASC GO No. 4, 9 Dec 1941.

2. ASC GO No. 4, 22 Jan 1943.

3. ASC GO No. 134, 19 Nov 1942.

4. AAF Reg. No. 20–4, 17 Oct 1941.

5. Interview with Maj Gen E. E. Adler, 16 Feb 1945.

6. Based on Memo: Col William H. Garrison to CG, ASC, "Service Groups," 26 Sep 1942, in TSHIS-2 files.

7. See section on "The Service Test at Fort Dix."

8. See AAF Regulations No. 15–127 and 15–128 in Supporting Documents.

9. "History of the Development of Fort Dix Development Center," Part II of unsigned, undated memo for Lt Gen Arnold in AAG Bulk # 104 (also in Brig Gen L. P. Whitten's file).

10. June 1942.

11. Col Joseph A. Mollison, CO, 2nd ASAC, to Chief, FSS, 24 Sep 1942, in TSPER 322 (Organization—Groups).

12. For detailed information, see the complete lists of Air Depot Groups and Service Groups in Appendix B, Supporting Documents.

13. Memo: Col William H. Garrison to CG, ASC, "Service Groups," 26 Sep 1942, in TSHIS-2 files, and interviews with Maj J. M. McCampbell and Maj R. W. Richardson, TSPTR.

14. The discussion of the 4th ASAC is based on interview with Mr. Norman Hamilton, TSPTR, 28 Feb 1946.

15. Memo: Col William H. Garrison to CG, ASC, "Service Groups," 26 Sep 1942, in TSHIS-2 files.

16. Col James T. Givens, Acting Chief, P&T Division, to CG, AAF, Attn: Director of Base Services, "Cadre Training," 30 Dec 1942, in TSAGD 353 (Training).

17. 1st Ind to *ibid:* Col James C. Cluck, for Brig Gen L. P. Whitten, Director of Base Services, to CG, ASC, 5 Jan 1943.

18. *Ibid.*
19. See section on "The McNarney Directive."
20. Interview with Maj J. M. McCampbell, TSPTR, 28 Feb 1946.
21. Thomas F. Brock, Adjutant General, to CG, AAF, "Assignment of Air Force Unassigned Enlisted Men Attending SOS Specialist Schools," 21 Nov 1942 (AG 220.31 (11–12–42) PE-A) and Lt Col T. M. Belshe for Col J. M. Bevans, Director of Personnel, to CG, ASC, "Procedure for Reporting and Distributing AAF Unassigned (ASWAAF) Enlisted Personnel," 24 Nov 1942, in TSAGD 353 (Training).
22. *Ibid.*
23. ASC Memo No. 50–2, 20 Aug 1942.
24. Ltr to CG, AAF, signed by Brig Gen E. E. Adler, but not mailed, 6 Feb 1943, in TSAGD 322 (Air Depot Groups—General).
25. Maj Gen Davenport Johnson, Director of Military Requirements, to CG, ASC, "Service Group Training," 23 Dec 1942, in TSAGD 353 (Training).
26. Ltr to CG, AAF, signed by Brig Gen E. E. Adler, but not mailed, 6 Feb 1943, in TSAGD 322 (Air Depot Groups—General).
27. *Ibid.*
28. Col James C. Cluck for Col L. P. Whitten, Director of Base Services, to CG, FS, ASC, "Replacement Units for Overseas Service," 12 Oct 1942, in TSHIS-2 files.
29. *Ibid.*
30. *Ibid.*
31. See speech of Brig Gen Charles E. Thomas, Jr., at ASC Conference, 6 Oct 1943, in "The History of Warner Robins Air Service Command to 1 Jan 1944," Vol. II.
32. See speech by Brig Gen E. E. Adler in "Proceedings—ASC Theater Supply Conference, Patterson Field, Fairfield, Ohio, 1–4 Nov 1943."
33. Memo to Col D. B. Schannep from Mr. Norman Hamilton, "Training at Service Group Training Centers, Muroc and Pendleton," Dec 1942, in TSHIS-2 files.
34. AAF Reg. No. 65–1, 14 Aug 1942.
35. Maj Gen W. H. Frank to CG, AAF, "Training of ASC Organizations," 18 Feb 1943, in TSPER 353.02 (Training).
36. *Ibid.*
37. 1st Ind to *ibid.*, dated 17 Mar 1943.
38. For a discussion of the effects of this system see Memo for the Director, ATSC, by Brig Gen E. E. Adler, Chief, Management Control, "Field Visit to Hq San Antonio ATSC, 8 Jan 1945 to 13 Jan 1945," approved 17 Jan 1945, in TSCMC-1 files.
39. AAF Reg. No. 20–31, 22 Dec 1943.
40. WAR, 16 Mar 1945, in TSHIS-2 files. For further information see

"Proceedings of Conference re Proposed Discontinuance of the 4500th and 4501st AAF Base Units, Lakeland and Venice Army Air Fields, Held at Hq, Third Air Force, Tampa, Fla., 8 Mar 1945," in Revised Training Structure Folder, TSPTR-2.

41. Memo 50–4, "Second Phase Training for Service Group," Avon Park, Florida, 13 Jul 1944, in TSHIS-2 files.

42. Brig Gen Robert W. Harper, AC/AS, Training, to CG, ASC, "Training Structure—Air Service Command," 22 Dec 1943, in Appendix A, Supporting Documents.

43. Maj Leo C. Wilson, Chief, Military Personnel, Hq ASC, to Maj Walter E. Nichol, Wellston Air Depot, Macon, Georgia, 30 Jan 1942, cited in "The History of Warner Robins Air Service Command," Vol. II.

44. A study of the special orders was made in 1944 by Sgt Herbert Toff, Historian of Herbert Smart Airport, Macon, Georgia.

45. *Ibid.*

46. "The History of Warner Robins Air Service Command," Vol. II.

47. GO No. 29, Hq ASC, 8 Apr 1942, and GO No. 6, Hq WAD, 11 Apr 1942.

48. SO No. 86, Hq ASC, 12 Aug 1942.

49. "The History of Herbert Smart Airport to 1 Jan 1944."

50. Col Russel Scott, Wellston AD, Macon, Georgia, to The Adjutant General, Washington, D.C., 30 Sep 1942, cited in "The History of Warner Robins Air Service Command," Vol. II.

51. Memo for Gen McMullen from Col Leo H. Dawson, "Status of Air Depot Group Training," 30 Jan 1943, in TSPER 322 (Organization—Air Depot Groups).

52. *Ibid.*

53. *Ibid.*

54. *Ibid.*

55. *Ibid.*

56. This section is based on "The History of San Antonio Air Service Command from Inception to 1 February 1943," pp. 60–64, 88–93.

57. *Ibid.*

58. AAF Reg. No. 20–4A, 27 May 1942.

59. See "The History of San Antonio Air Service Command from Inception to 1 February 1943," p. 63.

60. *Ibid.*

61. GO No. 1, Hq SAAD, Duncan Field, 9 Jan 1943.

62. GO No. 29, Hq ASC, 8 Apr 1942, and GO No. 66, Hq ASC, 11 Jul 1942.

63. *Ibid.*

64. See "The History of San Antonio Air Service Command from Inception to 1 February 1943."

65. *Ibid.*

## CHAPTER 8

1. GO No. 134, Hq ASC, 19 Nov 1942.
2. GO No. 141, Hq ASC, 8 Dec 1942; 142, 12 Dec 1942.
3. ASC Reg. No. 20-2, 14 Jan 1943; GO No. 4, Hq ASC, 22 Jan 1943.
4. Brig Gen L. P. Whitten, Director of Base Services, to CG, ASC, "Training Program for Air Depot and Service Groups," 31 Jan 1943, in TSAGD 353 (Training).
5. Memo to Col Schannep from Brig Gen E. E. Adler, Chief of P&T Division, "Training Program," 3 Mar 1943, in *ibid.*
6. *Ibid.*
7. Memo for E. E. Adler, Brig Gen, USA, Chief, P&T Division, by Col D. B. Schannep, Chief, T&O Section, "Training Plan for Air Depot Groups," 3 Mar 1943, in *ibid.*
8. *Ibid.*
9. *Ibid.*
10. R&R, Col J. C. Gordon, Asst Chief, P&T Division, to Chief, Program Planning, 19 Apr 1943, in TSAGD 353 (Training) and WAR, Hq ASC, 23 Apr 1943.
11. Memo for E. E. Adler, Brig Gen, USA, Chief, P&T Division, by Col D. B. Schannep, Chief, T&O Section, "Training Plan for Air Depot Groups," 3 Mar 1943, in TSAGD 353 (Training).
12. See "Record of Conference Held at HQ, ASC, Patterson Field, Fairfield, Ohio, April 9th and 10th, 1943," in personal file of Maj J. M. McCampbell, TSPTR, "History."
13. Maj Gen W. H. Frank to CG, AAF, "Training of A.S.C. Organizations," 18 Feb 1943, and 1st Ind thereto, 17 Mar 1943, in TSPER 353.02 (Training).
14. E. E. Adler for Maj Gen W. H. Frank to CG, AAF, "Revised Training Program for Service and Air Depot Groups," 19 Apr 1943, in TSAGD 353 (Training).
15. *Ibid.*
16. *Ibid.*
17. *Ibid.*
18. See remarks of Col D. B. Schannep in "Record of Conference Held at Hq, ASC, Patterson Field, Fairfield, Ohio, April 9th and 10th, 1943," in personal file of Maj J. M. McCampbell, TSPTR, "History."
19. 1st Ind (E. E. Adler for Maj Gen W. H. Frank to CG, AAF, "Revised Training Program for Service and Air Depot Groups," 19 Apr 1943), 30 Apr 1943, in TSAGD 353 (Training).
20. *Ibid.*
21. Another 1st Ind to *ibid:* Maj Gen O. P. Echols to AC/AS, Personnel, thru AC/AS, OC&R (Programs), 23 Apr 1943.
22. Maj Gen W. H. Frank to CG, AAF, "Activation of Parent Group (Re-

duced)," 17 Apr 1943 and 1st Ind thereto in TSAGD—Gen Frank file—353.02 (Training Program—Service Groups).

23. Exhibit No. 17, Appendix A, Supporting Documents.

24. Based on letter, Col D. B. Schannep, Chief, T&O Section, P&T Division, to CG's and CO's, Area ASC's, "Revised Training Program for Service and Air Depot Groups," 1 Jun 1943, in Appendix A, Supporting Documents.

25. *Ibid.*

26. *Ibid.*

27. Col Bernard Cummings, CO 303rd Service Group, Dale Mabry Field, Tallahassee, Florida, to CO Warner Robins ASC, "Revised Training Program—Service Groups," 17 Jul 1943, in TSAGD 353 (Training).

28. WAR, Hq ASC, 13 Aug 1943. Additional information on this visit was obtained in an interview with General Joseph T. McNarney, 7 April 1949.

29. Memo for the CG, AAF, from Lt Gen Joseph T. McNarney, Acting Chief of Staff, "Tactical Service Units to Support Combat Units," 13 Aug 1943, in Appendix A, Supporting Documents.

30. *Ibid.*

31. *Ibid.*

32. Exhibit No. 21, Appendix A, Supporting Documents.

33. WAR, Hq ASC, 27 Aug 1943.

34. Brig Gen Robert W. Harper, AC/AS, Training, to CG, 2nd Air Force, "Tactical Service Units to Support Combat Units," 16 Oct 1943, in TSHIS-2 files.

35. *Ibid.*

36. Maj Gen Barney M. Giles, Chief of the Air Staff, to CG, ASC, "Tactical Service Units to Support Combat Units," 25 Sep 1943, in TSHIS-2 files.

37. Exhibit No. 24, Appendix A, Supporting Documents.

38. Brig Gen E. E. Adler, Chief, P&T Division, ASC, to CG, AAF, "A Plan to Stabilize Air Service Command Military Personnel and Training Requirements," 16 Oct 1943, in Appendix A, Supporting Documents.

39. *Ibid.*

40. *Ibid.*

41. WAR, Hq ASC, 15 Oct 1943.

42. WAR, Hq ASC, 5 Nov 1943.

43. *Ibid.*

44. Brig Gen Robert W. Harper, AC/AS, Training, to CG, ASC, "Future Training Structure for Air Service Units," 10 Nov 1943, in Appendix A, Supporting Documents.

45. AAF Reg. No. 20–31, 22 Dec 1943 and 4 Feb 1944.

46. AAF Reg. No. 20–31, 31 Aug 1944.

47. Brig Gen E. E. Adler, Chief, P&T Division, to Comdrs, all area ASC's, "Tactical Service Units to Support Combat Units," 6 Jan 1944, in TSHIS-2 files.

48. Mimeographed ASC Letter Directive, "Administrative Plan for the Training of Service Groups," 16 Nov 1943, in *ibid.*

49. Memo, Lt Col Dallas M. Speer to Historical Section, "Training Air Service Units for Combat Theaters," 1 May 1944, in *ibid.*

50. Col W. W. Wood, Asst Chief, P&T Division, to CG's, Area and Overseas ASC's and Miami AD, "Projected Training Program," 11 Jan 1944, in Appendix A, Supporting Documents.

51. Interview with Maj Gen E. E. Adler, 16 Feb 1945.

52. ASC Memo No. 20–1, 16 Dec 1942.

53. Memo for Col Taylor from H. E. J. (Maj James, Control Room Officer) in TSAGD 353 (Training).

54. R&R from ASCCO to ASCPD, 17 Mar 1943, in *ibid.*

55. Col James G. Taylor, Control Officer, to ASCPT (Attn: Col Gordon), "Number of Depot Groups and Service Groups—World Wide, Actual, and Projected," 23 Apr 1943, in *ibid.*

56. Notes on Division Chiefs' Meeting, 2 Sep 1943.

57. WAR, Hq ASC, 21 May 1943.

58. ASC Bulletin 20–1, 15 Sep 1943.

59. Maj Gen W. H. Frank to Gen H. H. Arnold, "Dear Hap," 15 Nov 1943, in TSAGD 353 (Training).

60. See memo by Col Paul E. Ruestow, "Notes Taken at a Conference," 10 Sep 1943, in TSAGD—Gen Frank file—319.1 (Report—Col Ruestow—7th Air Force).

61. Col C. J. Bondley, Jr., CO, Puerto Rico Air Depot, to CG, ASC, Attn: Brig Gen Clements McMullen, "Comments on Overseas Movement," 14 Feb 1943, in AAG 322 D (Groups).

62. WAR, Hq ASC, 30 Jul 1943.

63. Notes on Division Chiefs' Meeting, 17 Jul 1943.

64. Notes on Division Chiefs' Meeting, 20 Nov 1943.

65. "History of Sacramento Air Depot, Jan 1942 to July 1943," p. 770.

66. Col Leo C. Wilson, Chief, Military Personnel Section, to CG's, Area ASC's, "Procurement Objective for Direct Commission," 7 Jun 1943, in TSPER 210.105 (Candidates for Commissions).

67. Notes on Division Chiefs' Meeting, 21 Jul 1943.

68. "History of the Officers Branch, Mil Pers Sec," memo in TSHIS-2 files.

69. Notes on Division Chiefs' Meeting, 18 Oct 1943.

70. WAR, Hq ASC, 25 Jun 1943.

71. WAR, Hq ASC, 9 Jul 1943.

72. "The History of San Antonio Air Service Command, 1 Feb 1943 to 1 July 1944," p. 96.

73. See Daily Diary, Air Services Division (Supply and Services), AC/AS, MM&D, 14 May 1943.

74. WD, AG 327.31 (5–20–43) PR-I, "Vol induction & enlistment of technicians for the ASC," 1 Jun 1943.

75. WAR, Hq ASC, 9 Jul, 6 and 13 Aug 1943.

76. *Ibid.*

77. Memo from Brig Gen E. E. Adler, ASCPD, to ASCPT (copy to Gen Frank), "Readiness of Units for Overseas Movements," 17 May 1943, in TSPER 353.01 (Training).

78. Col T. C. Odum, Deputy, The Air Inspector, to Maj Gen W. H. Frank, "POM Reports Indicating Lack of Supervision and Inspection by Higher Command of ASC Units Alerted for Overseas Movement," 19 Jun 1943, and 1st Ind thereto (9 pages long), in TSAGD—Gen Frank file—333 (Inspection Prior to Overseas Duty).

79. Memo for Col Jess B. Bennett from Lt Col Linn S. Chaplin, Asst Chief, T&O Section, "Deactivation of Certain Air Service Command Service Groups and Depot Groups," 23 Jun 1943, in TSPER 353.01 (Training).

80. "Inspection of the Air Service Command," Report of Brig Gen Junius W. Jones, the Air Inspector, to CG, AAF, 1 Sep 1943, in TSAIR 333.1 (Annual General Inspections), hereafter cited as "Annual General Inspection by Brig Gen Junius W. Jones."

81. "The History of San Antonio Air Service Command, 1 Feb 1943 to 1 July 1944," p. 77.

82. "Annual General Inspection by Brig Gen Junius W. Jones."

83. Condensed from memo for Col Bennett by Maj James G. Badgett, "Factors Contributing to Delay and Confusion in the Training of Service Groups," 22 Jul 1943, in TSPER 353.01 (Training).

84. On 21 Jul 1943 the 16th, 30th, and the 32nd Air Depot Groups, the 84th, 85th, 89th, 312th, 313th, and 316th Depot Repair Squadrons, the 57th, 59th, 71st, 74th, 84th, and 97th Service Groups, and the 383rd and 389th Aviation Squadrons were listed by Brig Gen L. P. Whitten as having been unfavorably reported on by the Air Inspector. In TSAGD—Gen Frank file—333 (Inspection Prior to Overseas Duty).

85. Memo from Brig Gen E. E. Adler, ASCPD, to ASCPT (copy to Gen Frank), "Readiness of Units for Overseas Movements," 17 May 1943, in *ibid.*

86. Gen H. H. Arnold to Maj Gen Walter H. Frank, "Lack of Readiness of Service Units Being Dispatched to Ports of Embarkation," 10 Aug 1943, in *ibid.*

87. Brig Gen E. E. Adler to CG, AAF, Attn: AC/AS, MM&D, 27 Jul 1943, in *ibid.*

88. *Ibid.*

89. WAR, Hq ASC, 18 Jun 1943.

90. *Ibid.*

91. 1st Ind, Brig Gen E. E. Adler to CG, AAF, Attn: AC/AS, MM&D, 27 Jul 1943, in TSAGD—Gen Frank file—333 (Inspection Prior to Overseas Duty).

92. Brig Gen Lucas V. Beau to CG, ASC, "Importance of Training of Depot Groups and Service Groups," 29 May 1943, in TSAGD—Gen Frank file—353 (Training).

93. See memo, Brig Gen C. W. Howard to CG, ASC, "Activation of Service Groups," 5 Aug 1943, in TSAGD 322 (Organization—Groups).

94. 1st Ind to *ibid.*, 30 Aug 1943.
95. "Annual General Inspection by Brig Gen Junius W. Jones."
96. *Ibid.*
97. *Ibid.*
98. WAR, Hq ASC, 27 Aug 1943.
99. WAR, Hq ASC, 13 Aug 1943 and 27 Aug 1943; also TSAGD 322 (Organization—Battalions).
100. Daily Diary, Air Services Division (Supply and Services), AC/AS, MM&D, 30 Jun 1943.
101. See AAF Regs. No. 15–127, 15–128, and 15–131 in Appendix A, Supporting Documents.
102. Memo for the Chief of the Air Staff by Brig Gen L. P. Whitten, Chief, Supply & Services Division, "Serv Units in AAF" (Gen Whitten file—Organization & Training).
103. "Annual General Inspection by Brig Gen Junius W. Jones."
104. *Ibid.*
105. *Ibid.*
106. WAR, Hq ASC, 20 Aug 1943.
107. "Annual General Inspection by Brig Gen Junius W. Jones."
108. *Ibid.*
109. *Ibid.*
110. Memo from Brig Gen E. E. Adler, ASCPD, to ASCPT (copy to Gen Frank), "Readiness of Units for Overseas Movements," 17 May 1943, in TSPER 353.01 (Training).
111. "Annual General Inspection by Brig Gen Junius W. Jones."
112. *Ibid.*
113. *Ibid.*
114. ASC Reg. No. 20–3, 19 Oct 1943, in Appendix A, Supporting Documents.
115. ASC Reg. No. 20–3, 28 Dec 1943, in *ibid.*
116. WAR, Hq ASC, 10 Sep 1943.
117. WAR, Hq ASC, 1 Oct 1943.
118. See ASC Reg. No. 50–2, 13 Oct 1943, in Appendix A, Supporting Documents.
119. "Annual General Inspection by Brig Gen Junius W. Jones."
120. ASC Regs. No. 20–3, 19 Oct 1943 and 28 Dec 1943, both of which are given in Appendix A, Supporting Documents.
121. "The History of San Antonio Air Service Command, 1 Feb 1943 to 1 July 1944," pp. 78–9.
122. Especially Gen Harper and Gen Whitten. Daily Diary, Air Services Division (Supply & Services), AC/AS, MM&D, 5 Jul 1943.
123. Conference of ASC Officers with the Air Staff, Hq AAF, 17 Aug 1943; in personal file of Maj J. M. McCampbell, TSPTR, "History."
124. *Ibid.*
125. *Ibid.*

126. *Ibid.*

127. *Ibid.* The discussion of General McNarney's visit is also based on an interview with the General, 7 April 1949.

128. Memo for the Commanding General, Army Air Forces, from Lt Gen Joseph T. McNarney, Acting Chief of Staff, "Tactical Service Units to Support Combat Units," 13 Aug 1943, in Appendix A, Supporting Documents.

129. Same as No. 123, above.

130. See Six Month Troop Forecasts or Commitment Charts, issued by AC/AS, OC&R; also see Complete Lists of Air Depot and Service Groups in Appendix B, Supporting Documents.

131. See WAR, Hq ASC, 25 Jun 1943 and subsequent WAR's. For information on the need for additional Service Squadrons for the ATC see Daily Diary, Air Services Division (Supply & Services), AC/AS, MM&D, 10 May, 8 Jun, 28 Jun, 22 Jul 1943.

132. See speech by Brig Gen E. E. Adler in "Proceedings—ASC Theater Supply Conference, Patterson Field, Fairfield, Ohio, 1–4 Nov 1943."

133. For further details see AAG (Gen Whitten file—Determination of Requirements for Service Units).

134. Same as No. 79, above.

135. Memo for Col D. B. Schannep, "Report of Trip to Hq, AAF, 11–12 Oct 1943," by Capt Ray W. Richardson and Lt Ingalls, in TSPER file (The Bradley Plan).

136. Same as No. 128, above.

137. WAR, Hq ASC, 9 Jul 1943.

138. "The Bradley Plan for the United Kingdom," Statistical Control Division, Management Control, Washington, D.C., now in TSHIS-2 files.

139. ATSC Historical Monograph, "The History of Estimating Requirements for AAF Organizational Equipment," p. 76.

140. WAR, Hq ASC, 13 Aug 1943.

141. Interview with Maj J. M. McCampbell and Mr. Norman Hamilton, TSPTR, 28 Feb 1946.

142. Memo by Brig Gen H. A. Craig, AC/AS, OC&R, to Gen Echols, "Implementation of the Bradley Plan," 10 Sep 1943, in TSSOV-6 files.

143. Same as No. 138, above.

144. TWX, Operations Division, WDGS: OPD 320.2 ETO (29 Aug 1943), 1 Sep 1943, in Logistics Planning Branch files, Hq AAF.

145. Memo by Brig Gen H. A. Craig, AC/AS, OC&R, to Gen Echols, "Implementation of the Bradley Plan," 18 Sep 1943, in TSSOV-6 files.

146. *Ibid.*

147. *Ibid.*

148. Memo by Brig Gen L. P. Whitten for AC/AS, OC&R, 9 Sep 1943, in TSPER file (The Bradley Plan).

149. *Ibid.*

150. Transcript of telephone conversation: Brig Gen L. P. Whitten and Brig Gen E. E. Adler, 10 Sep 1943, in *ibid.*

151. Memo for Col D. B. Schannep, "Report of Trip to Hq, AAF, 11–12 Oct 1943," by Capt Ray W. Richardson and Lt Ingalls, in *ibid.*

152. Same as No. 132, above.

153. WAR, Hq ASC, 17 Sep 1943.

154. WD File AG 370.5 (7–12–43) OB-S-E-GN-AF-SPMOT-M, 1 Aug 1943, short title "POM," and WD File AG 370.5 (8–6–43) OB-S-AF-M, 1 Aug 1943, short title "Air POM."

155. WAR, Hq ASC, 20 Aug 1943.

156. TWX, ASCSUP-BK-33, "Shipment of Personnel," 8 Nov 1943, in TSHIS-2 files.

157. Ltr Hq AAF, "Movement Orders, Shipment AG 870" (AAF 370.5, 14 Sep 1943, PUB-R-AF-M), dated 15 Sep 1943, as amended 17 Sep 1943.

158. WAR, Hq ASC, 17 Sep 1943, 24 Sep 1943, 1 Oct 1943, 8 Oct 1943, 29 Oct 1943, 5 Nov 1943, and 3 Dec 1943.

159. WAR, Hq ASC, 4 Nov 1943.

160. For a more detailed discussion of the subject of Organizational Equipment see ATSC Historical Monograph, "Supply of AAF Organizational Equipment to the Army Air Forces."

161. See ATSC Historical Monograph, "Supply of AAF Organizational Equipment to the Army Air Forces," p. 22.

162. That the system failed in many respects is indicated by the fact that as late as 21 Jul 1943 the 16th, 30th, and 32nd Air Depot Groups, the 84th, 85th, 89th, 312th, 313th, and 316th Depot Repair Squadrons, the 57th, 59th, 71st, 74th, 84th, and 97th Service Groups, and the 383rd and 389th Aviation Squadrons were listed by Brig Gen L. P. Whitten as having been unfavorably reported on by the Air Inspector for shortages of equipment. In TSAGD—Gen Frank file—333 (Inspection Prior to Overseas Duty).

163. WD File AG 370.5 (7–12–43) OB-S-E-GN-AF-SPMOT-M, 1 Aug 1943, short title "POM."

164. *Ibid.*

165. Maj Gen W. H. Frank to CG, AAF, "Supply of Air Force Organizational Equipment," 4 May 1943, in TSAGD—Gen Frank file—401 (System of Supply).

166. *Ibid.*

167. See remarks by Col D. B. Schannep in "Record of Conference Held at Hq, ASC, Patterson Field, Fairfield, Ohio, April 9th and 10th, 1943," in personal file of Maj J. M. McCampbell, TSPTR, "History."

168. Brig Gen E. E. Adler to CG, AAF, AC/AS, MM&D, Attn: Col Paul E. Ruestow, "Air Depot Group Training," 19 Apr 1943, in TSAGD—Gen Frank file—353.02 (Training Program—Service Groups).

169. Col Paul E. Ruestow to The Air Inspector, "Air Depot Training," 27 May 1943, in *ibid.*

170. Brig Gen B. E. Meyers, Dep AC/AS, MM&D, to CG, ASC, "Unsatisfactory Final Status Reports," 16 Apr 1943 and 1st Ind thereto: Col H. J. Knerr, Asst to CG, ASC, to CG, AAF, AC/AS, MM&D, 22 Apr 1943, in *ibid.*

171. Letter from the 68th Service Group forwarded to Col D. B. Schannep, Chief T&O Section, ASC, by Maj C. A. States, Chief P&T Division, Sacramento ASC, in TSPER 353.01 (Training Program).

172. Col Leo H. Dawson to Brig Gen Clements McMullen, Hq ASC, 19 Jan 1943, in TSAGD 320.3 (Tables of Organization).

173. See AGO Memo No. 700–17–43, 26 Mar 1943, in Appendix A, Supporting Documents, and AAF Reg. 65–50, 15 Apr 1943 and 15 Dec 1943.

174. Maj Gen W. H. Frank to CG, AAF, "Supply of Air Force Organizational Equipment," 4 May 1943, in TSAGD—Gen Frank file—401 (System of Supply).

175. "Development of the Air Force Logistical System."

176. See remarks by Capt H. A. Mullen in "Record of Conference Held at Hq, ASC, Patterson Field, Fairfield, Ohio, April 9th and 10th, 1943," in personal file of Maj J. M. McCampbell, TSPTR, "History."

177. *Ibid.*

178 *Ibid.* Engine overhaul equipment was covered by T.O. 00–30–105.

179. Same as No. 174, above.

180. Interview with Maj J. M. McCampbell, TSPTR, 22 Feb 1945. WD T/A 1–26, "AAF Operational Training Stations, Air Service Command Units," 21 Sep 1943, prescribed OTU equipment authorized at each station engaged in OTU training for ASC units.

181. ASC Reg. No. 65–29, 6 Feb 1943. For further information, see section on Shipment of Organizational Equipment and Supplies in ATSC Historical Monograph, "AAF Supply of Overseas Air Forces."

182. Memo for Brig Gen E. E. Adler, Chief P&T Division, "NYASPAC, Newark, New Jersey," 2 Apr 1943, in TSAGD—Gen Frank file—333.1 (Inspection Trips).

183. 1st Ind (Hq AAF to Hq ASC, "Procedure Covering Preshipment of Organization Equipment," 6 Aug 1943), Overseas Supply Section, ASC, to AC/AS, MM&D, 10 Aug 1943, in TSSOV-6 file (Pre-shipments).

184. Maj L. A. Shelton, Asst, Overseas Supply Section, to Col W. D. Dana, Chief, Overseas Supply Section, % 8th Air Force Service Command, London, England, 30 Sep 1943, in TSSOV-6 file.

185. WAR, Hq ASC, 24 Sep 1943.

186. For further information see section on Implementation of the Bradley Plan in ATSC Historical Monograph, "The History of Estimating Requirements for AAF Organizational Equipment."

187. WAR, Hq ASC, 7 Jan 1944.

188. Col Joseph T. Morris, Asst Chief, Maintenance Division, to ASCPD, through ASCSD, 10 May 1943, in TSAGD 353 (Training).

189. *Ibid.*

CHAPTER 9

1. GO No. 10, Hq AAF, 17 Jul 1944 and AAF Reg. No. 20–43, 17 Jul 1944.
2. *Ibid.*
3. ATSC Reg. No. 20–1, 1 Sep 1944.
4. ATSC Reg. No. 20–17, 1 Sep 1944.
5. Interview with Col Dallas M. Speer, TSAIR, 26 Mar 1946.
6. *Ibid.* Also interview with Maj J. M. McCampbell, TSPTR.
7. *Ibid.*
8. See "Report on Conference Held in Washington to Provide Service Units for CBI Theater, 1300, 28 Dec 1943," by Lt Col Kenneth H. Havens, Maj J. M. McCampbell, and Lt Robert E. Ingalls, in TSPER 352.02 (Training).
9. Interview with Col Dallas M. Speer, TSAIR, 26 Mar 1946.
10. *Ibid.*
11. Daily Diary, Air Services Division (Supply & Services), AC/AS, MM&D, 24 May 1944, 3 Jun 1944. See also the correspondence on the Air Commando—Combat Cargo Project, May–Jun 1944, in Logistics Planning Branch files, AC/AS, MM&D.
12. See memo to Gen Adler from Col Leslie H. Ross and Maj J. M. McCampbell, TSPTR, "Attendance at Conference, Hq, III Air Force, Tampa, Fla, 16 May 1944, and Report of Training Trip through Warner Robins Area," 22 May 1944, in personal file of Maj J. M. McCampbell, TSPTR (Memorandums—1944).
13. Interview with Col Dallas M. Speer, TSAIR, 26 Mar 1946.
14. *Ibid.*
15. *Ibid.*
16. For further information see "History of SATSC Training, Air Depot Groups, Service Groups, Miscellaneous Units," memo in TSHIS-2 files.
17. Interview with Col Dallas M. Speer, TSAIR, 26 Mar 1946.
18. See "Report on Training at Certain Western Areas Visited by Capt Walker, January and February 1945," memo in TSHIS-2 files.
19. *Ibid.*
20. *Ibid.*
21. *Ibid.*
22. Memo for Brig Gen Lucas V. Beau by Col R. M. Batterson, Chief, Military Training Section, "Jet-Propelled Training," 19 Oct 1944, in personal file of Maj J. M. McCampbell, TSPTR (Memorandums—1944).
23. *Ibid.*
24. *Ibid.*
25. Memo for Col Wood by Maj J. M. McCampbell, "Jet-Propelled Service Group," 21 Feb 1945, in personal file of Maj J. M. McCampbell (Memoranda —1945).
26. Unit Historical File, TSCSU-2a.
27. "Depot Unit, Army, Program at Ogden, ATSC. Observations by Capt F. D. Walker on visit 2 Feb 1945," a memo in TSHIS-2 files.

28. Brig Gen E. E. Adler, Chief, P&T Division, to CG, Ogden ASC, "Training of 1st and 2nd Depot Units, Army," 8 Jul 1944, Appendix A to "Training Report of Depot Units, Army, OASC," 7 Jul 1945.

29. *Ibid.*

30. See remarks of Capt W. E. Mineck, Hq AGF, in minutes of conference held at Hq ASC, 28 Mar 1944, in TSHIS-2 files.

31. See remarks of Lt Col O. N. Pratt in *ibid.*

32. Minutes of these conferences available in TSHIS-2 files.

33. WD T/O&E 1–407.

34. 1st & 2nd were activated by GO No. 30, Hq Ogden ATSC, 27 Jun 1944; 7th & 8th by GO No. 41, Hq Ogden ATSC, 7 Aug 1944; and the 9th on 13 Apr 1945.

35. AAF Training Standard No. 40–24.

36. "Training Report of Depot Units, Army, OATSC," 7 Jul 1945.

37. *Ibid.*

38. *Ibid.*

39. Change 1, T/O&E 1–407.

40. "Training Report of Depot Units, Army, OATSC," 7 Jul 1945.

41. See correspondence between Maj Gen Robert W. Harper, AC/AS, Training, and Brig Gen E. E. Adler, Chief, P&T Division, ATSC, in AFACT-4 files.

42. This section is based on ATSC Historical Monograph "A History of the Army Aircraft Repair Ship Project, November 1943–September 1944," which contains a detailed treatment of the whole AARSP. All information not otherwise acknowledged was obtained from this monograph. The purpose of this section is to present a brief synopsis of the project to indicate its relationship to the over-all program.

43. See Aircraft Repair and Maintenance Units (Floating) Training Manual in Appendix C, Supporting Documents.

44. DAR, AC/AS, MM&D, 23 May 1944.

45. Col L. P. Whitten, Director of Base Services, to Chief, TC, "AAF Floating Construction Requirements, FY 1944," 23 Nov 1942, in AAG 560E.

46. Maj Gen Walter H. Frank, CG, ASC, to CG, AAF, "Overhaul of Engines in the Southwest Pacific," 5 Mar 1943, in TSAGD—Gen Frank file—452.031.

47. Interview: Capt Roger H. McDonough with Col Paul E. Ruestow, Chief, Logistics Planning Branch, AC/AS, MM&D, 30 Sep 1944.

48. *Ibid.*

49. Maj Gen O. P. Echols, AC/AS, MM&D, to CG, ASC, "Floating Fourth Echelon Maintenance Facilities," 12 Nov 1943, in TSAGD 560.

50. 1st Ind (Maj Gen O. P. Echols to CG, ASC, 12 Nov 1943) Maj Gen Delmar H. Dunton to CG, AAF, 1 Dec 1943, in *ibid.*

51. Memo, Maj Gen O. P. Echols, AC/AS, MM&D, to Chief, AS, 11 Dec 1943, in AAG Bulk # 323.3.

52. Interview: Capt Roger H. McDonough with Maj Miles Kracman, TSMCO, 10 Nov 1944.

53. JCS 669/2, 25 Jan 1944, pp. 5–6. Also see memo, Admiral E. J. King, Commander in Chief, U.S. Fleet, Navy Department, to CG, AAF, 31 Jan 1944, in AAG 560F.

54. Memo, Brig Gen L. P. Whitten, Chief, Air Services Division, AC/AS, MM&D, to CG, AAF, 27 Jan 1944, in personal file of Maj David D. Lent, Office of the Marine Section, Hq AAF.

55. Admiral William D. Leahy, USN, for the JCS, to Rear Admiral E. S. Land, Chairman, U.S. Maritime Commission, dated 17 Jan 1944, mailed 8 Feb 1944, in JCS 669, Maj Lent's file.

56. Memo, Maj Gen L. D. Clay, Director of Materiel, ASF, to Chief, TC, 26 Apr 1944, in *ibid.*

57. See "Conversion of Repair Ships" in Chapter II, ATSC Historical Monograph, "A History of the Army Aircraft Repair Ship Project, November 1943—September 1944."

58. See "Conversion of Auxiliary Vessels," in *ibid.*

59. See "Shops and Equipment," in *ibid.*

60. See "Acquisition of Supplies," in *ibid.*

61. See "Armament for Repair Ships and Auxiliaries," in *ibid.*

62. See Chapter III, Personnel, in *ibid.*

63. Daily Diary, Air Services Division (Supply & Services), AC/AS, MM&D, 2 Mar 1944.

64. WAR, Hq ASC, 9 Feb 1945.

65. See Organization Charts in Aircraft Repair and Maintenance Units (Floating) Training Manual in Appendix C, Supporting Documents.

66. See WD T/O&E 1-911.

67. See AAF Training Standard 40–12, 29 May 1944.

68. See Chapter IV, Training, in ATSC Historical Monograph, "A History of the Army Aircraft Repair Ship Project, November 1943—September 1944."

69. WAR, Hq ASC, 9 Feb 1945.

70. *Ibid.*

71. This section is based on "The History of the 14th Service Group (Chinese), Venice Army Air Field, Venice, Florida," prepared by S/Sgt John S. Stuart, 15 Sep 1944. All information not otherwise acknowledged was taken from this history and from interviews with the Group's Commanding Officer, Col Eric H. Kaeppel, and with Sgt Stuart.

72. See manual in Appendix C, Supporting Documents.

73. *Ibid.*

74. Interview with S/Sgt John S. Stuart at Warner Robins, 15 Sep 1944.

75. "The History of the 14th Service Group (Chinese), Venice Army Air Field, Venice, Florida."

76. WD, MR 1-2, 15 Jul 1939, in Appendix A, Supporting Documents.

77. Par 2, a (1) in *ibid*. Caps and italics in the original.

78. Par 2, a (3) in *ibid*.

79. Par 13, b in *ibid*.

80. Memo, "War Department Policy in Regard to Negroes," attached to R&R, AC/AS, OC&R, to Troop Basis Division, 5 May 1944, in Colored Units file of Troop Basis Division, AC/AS, OC&R.

81. *Ibid*.

82. WD AG 210.31 (3 Jan 1944) OB-S-AM, 7 Jan 1944, in Command Subject file, Military Personnel Section, P&BS Division, ATSC.

83. Maj Gen George E. Stratemeyer, C/AS to AC/AS, MM&D, 20 Jun 1943, in TSPER 353.01 (Training Program).

84. *Ibid*.

85. WAR, Hq ASC, 27 Aug 1943.

86. Daily Diary, Air Services Division (Supply & Services), AC/AS, MM&D, 24 Sep 1943.

87. Brig Gen E. E. Adler, Chief, P&T Division, to CG, 1st Air Force, Mitchel Field, Hempstead, New York, "96th Service Group (Colored) at Oscoda, Michigan," 23 Aug 1943, in TSAGD 353 (Training).

88. *Ibid*.

89. Daily Diary, Air Services Division (Supply & Services), AC/AS, MM&D, 6 May 1944.

90. Brig Gen E. E. Adler, Chief, P&T Division, to CG, AAF, "Report of Investigation of the 96th Service Group (Colored), Oscoda, Michigan," 11 Aug 1943, in TSAGD 333.1 (Inspection of AAF Units. Aug thru Dec 1943).

91. *Ibid*.

92. 1st Ind to *ibid*, Brig Gen Edwin S. Perrin, DC/AS, to CG, ASC, 29 Aug 1943.

93. Memo for Col Woods from Maj James G. Badgett, Chief, Training Plans Unit, "96th Service Group," 3 Sep 1943, in TSAGD 322 (Groups).

94. Memo for Col J. C. Gordon, ASCPD, from Maj Earl H. Devanney, Military Personnel Section, "Narrative Report on Oscoda, Michigan Visit," 21 Jul 1943, in TSAGD 322 (Service Groups).

95. *Ibid*.

96. *Ibid*.

97. Col Leslie H. Ross to CO, Fairfield ASC, 10 Aug 1943, in TSAGD 322 (Groups).

98. *Ibid*.

99. *Ibid*.

100. Same as No. 94, above.

101. See Complete List of Service Groups in Appendix B, Supporting Documents.

102. *Ibid*.

103. "387th Service Group, Daniel Field, Georgia, Historical Data, 1 October 1944 to 31 October 1944," by Melburne House, Capt AC, Group S–3.

104. Memo for TSDEP Brig Gen Lucas V. Beau, Chief, P&BS Division, "Recommendation Relative to the 387th Air Service Group (Colored)," 20 Jan 1945, in personal file of Maj J. M. McCampbell, TSPTR (Memoranda—1945).

105. "387th Service Group, Daniel Field, Georgia, Historical Data, 1 December 1944 to 31 December 1944," by Melburne House, Capt AC, Group S-3.

106. *Ibid.*

107. *Ibid.*

108. *Ibid.*

109. Memo for TSPER from Maj J. M. McCampbell, "Report of Trip," 2 Jan 1945, in personal file of Maj J. M. McCampbell, TSPTR (Memoranda—1945).

110. Same as No. 105, above.

111. *Ibid.*

112. See Aviation Squadron Training Manual in Appendix C, Supporting Documents.

113. "Squadron History of 473rd Aviation Squadron from 28 December 1944 to 5 March 1945."

114. Similar statements were made to the writer by several colored soldiers assigned to Aviation Squadrons at Patterson Field, Ohio.

115. AAF Training Standard No. 40–14, 2 Oct 1943.

116. See "History of the 472nd Aviation Squadron, San Bernardino Army Air Field, California," 1 Apr 1945 to 19 Jun 1945.

117. Brig Gen Lucas V. Beau to CG, Fairfield ASC, "Reduction in Venereal Disease Rates," 18 Feb 1944, in TSAGD—Gen Frank file—726.1 (Sex Hygiene—Venereal Disease).

118. "History of the 456th Aviation Squadron, from Activation 12 February 1944 to 10 December 1944, San Bernardino Army Airfield, California."

119. Memo "ASC Survey of Negro Organizations," 10 Jul 1944, in TSAGD—Gen Frank file—322 (Troops—Negro).

120. Brig Gen L. P. Whitten, Office AC/AS, MM&D, to CG, ASC, "Training of Aviation Squadrons," 16 Aug 1944, in TSAGD—Gen Frank file—333.1 (Inspection—Fields and Depots).

121. See 2nd Ind, Brig Gen Paul C. Wilkins, CG, San Antonio ASC, to CG, ASC, 2 Aug 1943, in TSAGD—Gen Frank file—333 (Inspection Prior to Overseas Duty).

122. *Ibid.*

123. *Ibid.*

124. *Ibid.*

125. *Ibid.*

126. *Ibid.*

127. *Ibid.*

128. WAR, Hq ASC, 18 Jun 1943.

129. *Ibid.*

130. R&R, Lt Col W. H. Draham, Acting Chief, Military Training Branch, to ASCPT (Attn: Col Schannep), 20 May 1943, in TSPER 322 (Colored Units).

131. GO No. 34, Hq ASC, 19 Nov 1942.

132. In Annual Report, ATSC, 1944, p. 4.

133. Transcribed telephone conversation between Gen Frank and Gen Whitten, in AAG (Gen Whitten file—Letters to Gen Frank).

134. WAR, Hq ASC, 14 Jan 1944.

135. See, for example, WD AG 210.31 (3 Jan 1944) OB-S-AM, 7 Jan 1944, in Command Subject File, Military Personnel Section, P&BS Division, ATSC.

136. See Plate X, Control Room chart, in the original of this study. (Fig. VIII.)

137. *Ibid.*

138. *Ibid.*

## CHAPTER 10

1. For further details see "Report on Arm and Service Integration Program," 15 May 1945, submitted by Brig Gen Byron E. Gates, Chief, Management Control, to the Chief of Air Staff on 23 May 1945.

2. *Ibid.*

3. WD Cir No. 59, 2 Mar 1942. This statement was verified in an interview with General Joseph T. McNarney, 7 April 1949.

4. Same as No. 1, above.

5. *Ibid.*

6. Memo for Brig Gen E. S. Perrin by Lt Col R. E. S. Deichler, Chief, Organ. Planning Division, Management Control, "Arm and Service Integration," 27 Jan 1944, in AAG 321 (Integration of Arms and Services into the Army Air Forces).

7. Memo for the CG's, AGF, ASF, and AAF, "Review of Present Organizational Structure of the Army," 17 Aug 1943, in Logistics Planning Branch files, AC/AS, 4.

8. For further details see AAF Historical Study No. 28.

9. Memo for the Chief of Staff from CG's, AGF, ASF, and AAF, "Review of Present Organizational Structure of the Army," 10 Oct 1943, in Logistics Planning Branch files, AC/AS, 4.

10. "Report on Arm and Service Integration Program," 15 May 1945, submitted by Brig Gen Byron E. Gates, Chief, Management Control, to the Chief of Air Staff, on 23 May 1945. In TSHIS-2 files.

11. AAF Memo 20–18, 9 Nov 1943.

12. Memo for CG, AAF, "Review of Present Organizational Structure of the Army," 3 Dec 1943, in Logistics Planning Branch files, AC/AS, 4.

13. AAF Reg. No. 20–31, 22 Dec 1943.

14. For example, see the confidential reports to Gen Frank cited in the introduction to "The History of Warner Robins Air Service Command, 1 Jan to 1 July 1944," and other major unit histories.
15. Same as No. 10, above.
16. Change 13, AR 170–10, 11 Sep 1944.
17. WD Cir No. 388, 27 Sep 1944.
18. AAF Historical Study No. 28.
19. Quoted in "Report on Arm and Service Integration Program," 15 May 1945, submitted by Brig Gen Byron E. Gates, Chief, Management Control, to the Chief of Air Staff on 23 May 1945. In TSHIS-2 files.
20. AAF Reg. No. 35–57, 9 Apr 1945.
21. See the voluminous correspondence on the proposed revisions of AAF Reg. No. 65–1, 14 Aug 1942, in response to Gen McNarney's letter to the Theater Commanders on 1 Apr 1943. See also the reports and recommendations of the Steering Committee on Foreign Theaters, Hq ASC. Both in TSHIS-2 files.
22. Memo for the Chief of Air Staff by Brig Gen Byron E. Gates, Chief, Management Control, "Integration," 10 May 1944, in Logistics Planning Branch files, AC/AS, 4.
23. This section is based on ATSC Historical Monograph, "History of the Supply, Maintenance, and Training for the B-29." All information not otherwise acknowledged was obtained from this monograph. The purpose of this section is to present a brief synopsis of the project, to indicate its relationship to the over-all program.
24. See Introduction to *ibid.*
25. See chapter on Maintenance and Depot Modification in *ibid.*
26. See *ibid.*, p. 84.
27. Lt Col Dallas M. Speer, B-29 Project Officer, P&T Division, ASC, to ASPP, 24 Dec 1943, in Control Room files.
28. *Ibid.*
29. Interview with Maj J. M. McCampbell, TSPTR, 13 Apr 1946.
30. Daily Diary, Air Services Division (Supply & Services), AC/AS, MM&D, 22 Feb 1944.
31. See Service Group (New Type) Training Manual in Appendix C, Supporting Documents. Also see P&T Division Cir No. 50–100–15, 23 May 1944.
32. Brig Gen E. E. Adler, Chief, P&T Division, ASC, to CG, OCASC, "Training in the B-29 OTU," 27 Nov 1943, in Control Room files.
33. Interview with Lt Col R. Slough, TSPTR, 6 Mar 1945.
34. *Ibid.*
35. See ATSC Historical Monograph "History of the Supply, Maintenance, and Training for the B-29," p. 92.
36. *Ibid.*
37. DAR, AC/AS, OC&R, 18 Feb 1944.

38. See Service Group (New Type) Training Manual in Appendix C, Supporting Documents.

39. See "Training Program for Service Groups," in ATSC Historical Monograph, "History of the Supply, Maintenance, and Training for the B-29."

40. *Ibid.*

41. Col W. W. Wood for Brig Gen E. E. Adler, Chief, P&T Division, ASC, to CG, AAF, Attn: AC/AS, Training, "Equipment for AAF Base Units in the Air Service Command Training Structure," 21 Jul 1944, in AFACT-4 files.

42. See "Training at Oklahoma City," in ATSC Historical Monograph, "History of the Supply, Maintenance, and Training for the B-29."

43. Interview with Maj. J. M. McCampbell, TSPTR, 13 Apr 1946.

44. These units were reorganized by the 2nd Air Force to include 14 officers and 34 enlisted men. ("History, Maintenance and Supply Div, Hq, 2nd Air Force, 1 July 1944 to 12 Nov 1945," in TSHIS-2 files).

45. Interview with Col Louis A. Merrillat, Special Training Coordinator, ASC, 4 May 1945.

46. Same as No. 42, above.

47. See "B-29 Training School," in ATSC Historical Monograph, "History of the Supply, Maintenance, and Training for the B-29."

48. See "Air Depot Group Training," in *ibid.*

49. See "Summary of Units Trained," in *ibid.*

50. Brig Gen W. W. Welch, Office of AC/AS, Training, to Director, ATSC, 10 May 1945, in TSPTR-2 files.

51. Teletype, 2nd Air Force F 2039, Maj Gen R. B. Williams, CG, 2nd AF to Lt Gen S. C. Streett, 11 June 1945, in *ibid.*

52. *Ibid.*

53. See presentation accompanying the R-type T/O&E's of 5 March 1945, prepared by Logistics Planning Branch, AC/AS, MM&D (redesignated AC/AS, 4), in TSHIS-2 files.

54. *Ibid.*

55. *Ibid.*

56. Memo, Col R. M. Batterson, Jr, Chief, Military Training Sec, to TSPER, 15 June 1945, in *ibid.*

57. Manning and Training Directive (Effective V-J Day) in TSAGD 353.01 Training Program—Schedules and Directives—Aug 1945 thru Dec 1945.

58. *Ibid.*

★★★★★

# SOURCES

## Public Documents


Public Law 242, 41 *Stat.* 759 (1920)

Public Law 781, 76th Congress, 3rd Sess. Chapter 717, "2nd Supplemental National Defense Appropriation Act (1941)," approved 9 September 1940, 54 *Stat.* 873

Public Law 353, 77th Congress, 1st Sess. Chapter 591, "3rd Supplemental National Defense Appropriation Act (1942)," approved 17 December 1941, 55 *Stat.* 812


## Training Manuals

*The following training manuals were consulted in the preparation of this study. Copies may be found in TSHIS-2 files, except as otherwise noted.*


"Organization and Functions, Hq and Hq Squadron, Air Depot Groups," ASC Preliminary Instruction Manual, May 1942.

"Aircraft, Organization and Operation, Air Depot Group Repair Squadron," ASC Preliminary Instruction Manual, May 1942.

"Aircraft, Attached Arms and Services, Air Depot Group," ASC Preliminary Instruction Manual, May 1942.

"The Fort Dix Training Manual," July 1942.

"Organization and Training of an Air Depot Group, Patterson Field, Fairfield, Ohio," 15 September 1942 (revised to December 1st).

"The Service Center," AAF Field Manual 1–195, 26 September 1942.

IV ASAC Memo No. 50–6, Training Manual for Air Service Groups of the Fourth Air Service Area, December 1942.

"Aircraft, Supply, Air Depot Group," ASC Preliminary Instruction Manual, January 1943.

"Plan for Civilian Training," ASC, January 1943.

"The Service Group," Patterson Field, Fairfield, Ohio, January 1943.

"The Air Force in Theaters of Operations, Organizations, and Functions," an AAF Manual, May 1943, in Logistics Planning Branch, AC/AS, M&S, Hq AAF.

ASF Manual M3 (Training), "Courses of Instruction Given in Schools of the Army Service Forces," Hq ASF, November 1944.

"Handbook on German Military Forces," WD Technical Manual, TM-E 30–451, 15 March 1945.

AFTAD Training Devices Catalogue, June 1945.

Revision of FM 1–40, "Intelligence Procedure in Aviation Units," Report of the Army Air Forces Center, Orlando, Florida, 7 August 1945.

*Copies of the following manuals may be found, in the order listed, in the Supporting Documents of this study, Appendix C, Training Manuals, in AFIHI archives:*

Air Depot Group Training Manual, 30 August 1944.
Service Group Training Manual, 22 January 1944.
Service Group (New Type) Training Manual, 23 May 1944.
Aircraft Repair and Maintenance Units, Floating, Training Manual, 3 July 1944.
Classification Procedures in Air Service Command Installations, 1943.
Air Service Command Digest of Educational Requirements for Military Personnel Students, 1 January 1944.
Aviation Squadron Training Manual, 20 May 1944.
Air Service Command Soldier's Guide, 25 March 1944.
Air Forces Manual No. 50, "The Unit Gas Officer and Gas Noncommissioned Officer," 17 October 1944.
Chinese Translation of the Required Articles of War, 30 May 1944.
Fundamentals of Chinese Mandarin, 22 February 1944.
Camouflage Highlights, 17 December 1943.
Camouflage Training Manual for the Air Service Command, 1 January 1943, Revised 1 August 1943.

## Regulations, Orders, and Directives

*The following publications were issued by the War Department:*

Army Regulations
Bulletins
Circulars
General Orders
Memoranda
Mobilization Regulations
Protective Mobilization Plans
Tables of Allowances (T/A)
Tables of Basic Allowances (T/BA)
Tables of Equipment (T/E)
Tables of Organization (T/O)
Tables of Organization and Equipment (T/O&E)
The Adjutant General's Letters and Memos
 Warning Orders
 Movement Orders
  File AG 370.5 (7–12–43) OB-S-E-GN-AF-SPMOT-M, 1 August 1943, short title POM

File AG 370.5 (8–6–43) OB-S-AF-M, 1 August 1943, short title Air POM

*The following publications were issued by the Army Air Forces:*

Hq Office Instructions,
Memoranda,
Regulations,
Training Standards
Organizational Equipment Lists (OEL)
Technical Orders (T.O.)
AFTAD Bulletins

AAG Miscellaneous
Troop Forecasts, or Commitment Charts, AC/AS, OC&R
Materiel Division Policy Notices

"Index to AAF Tables of Organization and Equipment, Tables of Organization, Tables of Distribution, and Manning Tables," in TSHIS-2 files

*The following were issued by various air force headquarters:*

Air Corps Circulars

Air Service Circulars

ASC Bulletins,
General Orders,
Letter Directives and TWX Directives,.
Memoranda,
Regulations,
Special Orders,
Personnel and Training Division Circulars

FO Memos
MC Regulations

ATSC Letters,
Regulations

Second Air Force Memoranda and Letter Directives
Third Air Force Memoranda

OATSC General Orders
OCATSC General Orders,
Regulations and Letter Directives
SAASC (SAAD) General Orders
WRASC (WAD) General Orders and Special Orders,

**Regulations**

**40th Parent Service Group Memos**

## Historical Monographs

*Copies of the following AAF Historical Studies and Overseas Histories may be found in the Historical Office, Secretary of Air Staff, Hq AAF (AFIHI archives):*

No. 10, "Organization of the Army Air Arm, 1935–1943."
No. 25, "Organization of Military Aeronautics, 1907–1935."
No. 26, "Individual Training in Aircraft Maintenance in the Army Air Forces."
No. 28, "Development of Administrative Planning and Control in the AAF."
No. 29, "Summary of Air Action in the Philippine Islands and Netherland East Indies—Pearl Harbor to March 1943."
Draft "History of Air Service Command in the European Theater."
"History of the Eighth Air Force Service Command," and Supporting Documents.
Report of the Historical Division, AAF, to WDGS, G-2, on training in the AAF, chapter entitled "Training of Service and Maintenance Units in the AAF."

*Copies of the following ATSC Historical Monographs may be found either in the Historical Office, Secretary of Air Staff, Hq AAF (AFIHI archives), or in the Historical Section, Area "A" Hq ATSC (TSHIS-2 files).*

"A History of the Army Aircraft Repair Ship Project, November 1943–September 1944."
"The Air Service Command, an Administrative History, 1921–1944," and Supporting Documents. (Earlier titles had different dates.)
"Acquisition of Facilities for the Air Service Command."
"The History of Estimating Requirements for AAF Organizational Equipment," Parts I and II.
"AAF Supply of Overseas Air Forces."
"Supply of AAF Organizational Equipment to the Army Air Forces."
"History of the Procurement and Training of Civilian Personnel (Air Service Command)," Parts I, II, and III.
"A History of the Maintenance of Army Aircraft in the Continental United States, Part I—1921–1939."
"A History of Technical Publishing by the Air Service Command—Air Technical Service Command—1923–1944."
"Radio and Radar Supply, Maintenance, and Training by the Air Service Command."

"History of the Supply, Maintenance and Training for the B-29."
"The B-29."

### Special Studies

*The Special Studies listed below may be found in TSHIS-2 files, except as otherwise noted:*

"Annual Report," Chief of Air Service—1923.
"The History of Rockwell Field," Maj H. H. Arnold, Air Service, 1923.
"Air Force Administrative Organization and Procedure," 23 July 1941.
"Notes on Strength of German Air Force Units," in Germany 9000 (1940–41) file, A-2 Library, Hq AAF.
"Revue de l'Armée de l'Air," in Germany 9100 (1935–43) file, A-2 Library, Hq AAF.
"Personnel and Training Activities of the Ordnance Section from 1 June 1942 thru 1 November 1943," prepared by Lt Col Joseph H. Bryson, TSPTR.
"Training Guide, Signal Section, Air Service Command," 1st issue, November 1943; 2nd issue January 1943.
"Inspection of the Air Service Command," report of Brig Gen Junius W. Jones, The Air Inspector, to CG, AAF, 1 September 1943, in TSAIR 333.1 (Annual General Inspections).
"Report of Investigation of the Air Service Command: Annual General Inspection," 1 September 1943, in TSAIR files.
"Development of the Air Force Logistical System," a memo prepared in 1943 by Col D. R. Stinson, Assistant Chief, Supply and Services Division, AC/AS, MM&D.
"The Bradley Plan for the United Kingdom," Statistical Control Division, Management Control, Hq AAF.
"Arm and Service Integration," memo for Brig Gen E. S. Perrin by Lt Col R. E. S. Deichler, Chief, Organization Planning Division, Management Control, 27 January 1944, in AAG 321 (Integration of Arms and Services into the Army Air Forces).
"Annual Report of Instructional Property Branch," 30 November 1944.
"Annual Report," ATSC, 1944.
"Annual Report," ATSC Area Commands, 1944.
"Chemical Warfare Units with the ATSC," memo prepared by Maj A. S. DeRossett, TSCWS.
"Finance Sections, Air Depot and Air Service Groups," memo prepared by Miss E. Jones, TSBFO.
"Medical History, Air Service Command, 1942–1945," Section VIII (*sic*), "Training," prepared by Maj Richard J. Brightwell, TSPTR.
"Historical Report, Quartermaster Activities, ASC, ATSC," memo prepared by Lt Col R. Slough, 6 March 1945.
"Report on Arm and Service Integration Program," 15 May 1945, submitted

by Brig Gen Byron E. Gates, Chief, Management Control, to the Chief of Air Staff on 23 May 1945.

"History, Maintenance and Supply Division, Second Air Force, 1 July 1945 to 12 November 1945."

"History of Training Literature within the ATSC," memo prepared by Capt J. R. Henson, TSPTR, 21 February 1946.

"Training Aids in ATSC," memo prepared by Capt C. M. Jessup, Jr., TSPTR, 1 March 1946.

"Historical Organization File," Statistical Control Office Memo, in TSCSU files.

*The following Special Studies may be found, in the order listed, in the Supporting Documents of this study, Appendix B, Statistical Summaries, in AFIHI archives:*

Regular Air Corps Active Organizations (Continental U.S. Only), 1 December 1941

Hutments and CC Type Prefabricated Buildings for Mobile Air Depot Groups, 1942

Technical Training Courses for Officers, 1942–1945

Technical Training Courses for Enlisted Men, 1942–1945

Individual Technical Training—Supplementary List, 1942–1945

Officers Service Schools and Courses Utilized by ATSC Since 1 January 1945, 1 November 1945

Enlisted Service Schools and Courses Utilized by ATSC Since 1 January 1945, 1 November 1945

Complete List of Air Depot Groups (based on Headquarters Squadrons), 15 December 1944

Complete List of Service Groups (based on Headquarters Squadrons), 31 December 1944

## Conferences

"Area Commanders' Conference, Patterson Field, Fairfield, Ohio, 4–6 October 1943," in TSHIS-2 files.

"Conference of ASC Officers with the Air Staff, Hq AAF, 17 August 1943," in personal file of Maj J. M. McCampbell, TSPTR ("History").

"Panel Discussion by the Chief of Staff and Special Staff Representatives of the Air Service Command at AAFSAT, 27 September 1943," in TSHIS-2 files.

"Proceedings of ASC Theater Supply Conference, Patterson Field, Fairfield, Ohio, 1–4 November 1943," in TSHIS-2 files.

"Proceedings of Conference re Proposed Discontinuance of the 4500th and 4501st AAF Base Units, Lakeland and Venice Army Airfields, held at Hq, Third Air Force, Tampa, Florida, 8 March 1945," in TSHIS-2 files.

"Record of Conference Held at Hq, ASC, Patterson Field, Fairfield, Ohio,

April 9th and 10th, 1943," in personal file of Maj J. M. McCampbell, TSPTR ("History").

"Theater Administration, Air Force Presentation at a War Department Conference," 8 February 1944, by Brig Gen L. P. Whitten, Chief, Air Services Division, AC/AS, MM&D, now in TSHIS-2 files.

## MAJOR UNIT HISTORIES (ATSC)

*Major Unit Histories may be found either in the Historical Office, Secretary of Air Staff, Hq AAF (AFIHI archives), or in the Historical Section, Area "A," Hq ATSC (TSHIS-2 files).*

"History of Fairfield Air Depot: 1917–1943."

"The History of Oklahoma City Air Technical Service Command, October to December 1944."

"History of Sacramento Air Depot, January 1942 to July 1943."

"History of Sacramento Air Service Command," Installments I to III.

"History of San Antonio Air Depot from Inception to 1 February 1943."

"The History of San Antonio Air Service Command from 1 February 1943 to 1 July 1944."

"The History of Warner Robins Air Service Command to 1 January 1944," Volumes I and II (by the present author).

"The History of Warner Robins Air Service Command, 1 January to 1 July 1944" (by the present author).

## UNIT HISTORIES

*Unit Histories may be found in the Historical Office, Secretary of Air Staff, Hq AAF (AFIHI archives). Some, as noted, may also be found in the Historical Section, Area "A," Hq ATSC (TSHIS-2 files).*

"The History of Herbert Smart Airport to 1 January 1944," in TSHIS-2 files.

"History of Daniel Field from Activation to 1 January 1944."

Historical Record, Engineering Department, Albuquerque AAB, New Mexico, in TSHIS-2 files.

"Brief History of the 1st Air Depot Group, 1 January 1941 to 7 December 1941."

"Brief History of Hq. and Hq. Squadron, 1st Air Depot Group 1 January 1941 to 7 December 1941."

"Organization History of the Fourth Air Depot Group from 1 January to 31 December 1942."

"Organization History of the 26th Air Depot Group from Date of Activation to 1 September 1943."

"History of the 3rd Service Group, Activation Date to Overseas Date."

"Organizational Development, Northeast Air District and First Air Force, 19 November 1940 to 31 December 1943."

"History of the III EAUTC, 18 March 1943 to 1 May 1944."
"Historical Data, Radio Section, 4000th AAF Base Unit (Command)," in TSHIS-2 files.
"Historical Report, VHF Engineering Agency, 1 February 1944," in TSHIS-2 files.
"History of the 1st Radar and VHF Installation and Maintenance Unit, 31 August 1943 to 31 May 1944," in TSHIS-2 files.
"Training Report of Depot Units, Army, OATSC," 7 July 1945, in TSHIS-2 files.
"The History of the 14th Service Group (Chinese), Venice Army Airfield, Venice, Florida," prepared by S/Sgt John S. Stuart, 15 September 1944, in TSHIS-2 files.
"387th Service Group, Daniel Field, Georgia, Historical Data, 1 October 1944 to 31 October 1944 (also December)," by Melbourne House, Capt, AC, Group S-3, in TSHIS-2 files.
"Squadron History of 473rd Aviation Squadron from 28 December 1944 to 5 March 1945," in TSHIS-2 files.
"History of the 472nd Aviation Squadron San Bernardino Army Airfield, California, 1 April 1945 to 19 June 1945," in THSIS-2 files.
"History of the 456th Aviation Squadron, from Activation 12 February 1944 to 10 December 1944, San Bernardino Army Airfield, California," in TSHIS-2 files.

## Correspondence and Other Files

### HEADQUARTERS, AAF, WASHINGTON, D.C.

*This list includes both classified and unclassified files:*

AAG Bulk—Unprocessed
Bulk No. 65, 66, 90, 104, 172
161 (Contract Schools)
321 (Integration of Arms and Services into the Army Air Forces)
321H (Quartermaster)
322D (Groups)
353.9 (Training—General)
353.9C (Training Programs)
353.9E (Training—General)
560E, F
General Whitten File
(Determination of Requirements for Service Units)
(Letters to General Frank)
(Organization and Training)
AFACT-4 Files: 040.053; 300.67
A-2 Library, Hq AAF: Germany 9000 (1940–41); Germany 9100 (1935–43)

AFIHI Archives (Contains histories of all air forces, commands, and units in the AAF)

Also complete files:

Daily Diary, Air Services Division (Supply and Services), AC/AS, MM&D

Daily Activity Report (DAR), Air Quartermaster

Daily Activity Report (DAR), AC/AS, MM&D

Daily Activity Report (DAR), AC/AS, OC&R

AFDAE-1 Files, AC/AS, MM&D

Logistics Planning Branch Files, AC/AS, MM&D

Marine Section Files, AC/AS, MM&D

JCS 669/2

Unit Training Division Files, AC/AS, Training

Colored Units File, Troop Basis Division, AC/AS, OC&R

## HEADQUARTERS, ATSC, WRIGHT FIELD, OHIO

*This list includes both classified and unclassified files:*

TSAGD Microfilm Unit, 1941 Correspondence, Reel 13, Item 11, and Reel 14, Item 25.

320.3 (Tables of Organization)
322 (Air Depot Groups)
322 (Air Depot Groups—General)
322 (Groups)
322 (Organization—AAF Units)
322 (Organization—Air Depot Groups)
322 (Organization—Battalions)
322 (Organization—Groups)
323.6 (Organization—Air Service Command)
333.1 (Inspection of AAF Units—August thru December 1943)
352.1–4 (Maintenance and Operation of Aircraft)
353 (Training)
381 (Victory Program)
560

General Frank File

319.1 (Report—Col Ruestow—7th Air Force)
322 (Troops, Negro)
333 (Inspection Prior to Overseas Duty)
331.1 (Inspection—Fields and Depots)
333.1 (Inspection Trips)
353 (Training)
353.01 (Training Program)
353.02 (Training Program—Service Groups)
400.359 (Contract Schools)

401 (System of Supply)
452.031
726.1 (Sex Hygiene—Venereal Disease)

TSAIR Annual General Inspection files
TSBCO 400. 35 (1942)
TSCMC-1
TSCON 600.121 (General)
600.121 (Depot Groups)
600.121 (Depot Groups—General)
600.121 (Depot Groups—October 1941 to March 1942)
600.121 (Depot Groups—July–September 1942)
TSCSU-2 Unit Historical Files
TSHIS-2 Files (formerly TSCHI-2)
Daily Activity Report (DAR), Hq, ASC
Daily Journal, Hq, ASC
Notes on Division Chiefs' Meetings, Hq, ASC
Weekly Activity Report (WAR), Hq, ASC
TSMCO 323.3 (Air Depot Group, Fourth)
TSPER 210.105 (Candidates for Commissions)
210.105 (Appointment of Key Civilians in the AUS)
322 (Air Depot Groups—General)
322 (Colored Units)
322 (Organization—Air Depot Groups)
322 (Organization—Groups)
327.02 (Selective Service)
352.02 (Training)
353.01 (Training)
353.01 (Training Program)
353.02 (Training)
— (First Provisional Service Group, Fort Dix, N.J.)
— Officer's Branch, Policy Book
— Schools Unit, Policy Book
File (The Bradley Plan)
TSPMP 220.31 (cited as ASCPM2B)
(Command Subject File)
TSPTR Personal Files of Maj J. M. McCampbell, TSPTR
(History)
(Memoranda—1944)
(Memoranda—1945)
Personal Files of Mr. Norman Hamilton, TSPTR
TSPTR-2
(Revised Training Structure Folder)
TSSOV-6
Files
File (Pre-shipments)

Control Room Presentations and Files (includes Committee Reports), Hq, ATSC

## Books and Periodicals

*AAF, The Official Guide for the Army Air Forces*
Pocket Books, March 1944.

*Report on the Army, July 1, 1939 to June 30, 1943*
Biennial Reports of General George C. Marshall, Chief of Staff of the United States Army, for the Secretary of War. The Infantry Journal, 1943.

*Class Book, Class 44 D*
AAF OCS, Miami Beach, Florida.

*U.S. Naval Logistics in the Second World War*, by Duncan S. Ballantine, Princeton (1947).

*The Army Air Forces in World War II*, Vol. I, edited by Wesley Frank Craven and James Lea Cate, Chicago (1948).

*The Organization of Ground Combat Troops*, by Kent Roberts Greenfield, Robert R. Palmer, and Bell I. Wiley, first in the series *The United States Army in World War II*, Washington (1947).

## Interviews

* Adkinson, Capt Emory H., Assistant Adjutant General, Warner Robins ATSC.

† Adler, Maj Gen E. E., TSCMC, formerly Chief, P&T Division, Hq ASC.

Badanes, Lt Col E. B., TSSCO.

Baker, Lt Col T. H., Military Personnel Section, P&T Division.

Batterson, Col R. M., TSPTR.

* Bennett, Miss Pauline I., TSCSU-2A.

Brewster, Col Pete, TSAIR.

Brightwell, Maj Richard J., TSPTR.

Buchanan, Capt W. A., TSPTR, with Mr. Paul Angle.

Canning, Lt Col Roscoe, TSSDL.

† Cannon, Lt Col Vincent T., Planning Section, Logistics Planning Branch, Air Services Division, AC/AS, M&S.

DeRossett, Maj A. S., TSCWS.

Dodds, Capt W. C., TSMAC.

† Freathy, Maj A. E., TSMCO.

* Gall, Maj Jack K., TSPTR.

Griffin, Capt Ira, TSPTR.

* Hamilton, Mr. Norman, TSPTR.

† Hitchcock, Lt Col C. R., Office of Deputy for Services Training, Unit Training Division, AC/AS Training.

* Hoberman, Lt Maxwell G., Control Office Warner Robins Air Service Command.

† Howe, Col J. D., TSAIR.
Hughes, Lt Paul, TSPTR.
† Jessup, Capt C. M., Jr.
Jones, Miss E., TSBFO.
Julia, Capt Nancy, TSPMP.
Kaeppel, Col Eric H., CO, 14th Service Group, Venice, Florida.
Kossich, Sgt, TSPTR.
Kracman, Maj Miles, TSMCO, with Capt Roger H. McDonough.
* League, Maj J. B., TSPMP.
Marley, Mr. G. R., TSMCO.
* McCampbell, Maj J. M., TSPTR.
McNarney, Gen Joseph T., CG, AMC.
Merrillat, Col Louis A., formerly Chief, P&T Division, San Bernardino ASC, and Special Training Coordinator, Warner Robins ASC.
Miller, Maj W. R., TSPMP.
Myers, Mr. V. J., TSMCO.
† Reass, Lt Col J. L., TSPMP.
Richardson, Maj. R. W., TSPTR.
Rossoff, Maj Isidor, Chief, T&O Section, P&T Division, Warner Robins ASC.
Ruestow, Col Paul E., Chief, Logistics Planning Branch, Hq AAF, with Capt Roger H. McDonough.
† Scott, Col Russel, Deputy Commander, Warner Robins ASC.
* Slough, Lt Col R., TSPTR.
† Speer, Col Dallas M., TSAIR.
† Stuart, S/Sgt John, Historian, 14th (Chinese) Service Group, Venice, Florida.
Thomas, Brig Gen Chas. E., Jr., Commanding General, Warner Robins ASC.
† Toff, Sgt Herbert, Historian of Herbert Smart Airport, Macon, Georgia.
Webster, Maj J. L., Military Personnel Section, P&T Division.
Wormwood, Mr. G. R., TSSDL.

* These individuals were interviewed not once but many times—in some instances over a period of a year or two. The author is greatly indebted to them for both facts and interpretation.

† These individuals were interviewed half a dozen times or more, on subjects connected with this study, either by the author or by other members of the Historical Office, ATSC.

N.B. Most of the others on this list were also interviewed several times. All of them are to be thanked for their patience and their understanding of the problems of the historian.

## GLOSSARY OF ABBREVIATIONS

| | |
|---|---|
| A-1 | Air Staff, Personnel |
| A-2 | Air Staff, Intelligence |
| A-3 | Air Staff, Training and Operations |
| A-4 | Air Staff, Supply |
| AAB | Army Air Base |
| AAF | Army Air Forces (or Field) |
| AAFSAT | Army Air Forces, School of Applied Tactics, Orlando, Florida |
| AAFTAC | Army Air Forces, Tactical Center (later designation for AAFSAT) |
| AAFTAD | Army Air Forces, Training Aids Division |
| AAFTC | Army Air Forces Training Command |
| AAG | Air Adjutant General |
| AARS | Army Aircraft Repair Ship |
| AARS/A | Army Aircraft Repair Ship, Auxiliary |
| AARSP | Army Aircraft Repair Ship Project |
| AB | Air Base |
| AC | Air Corps |
| AC/AS | Assistant Chief, Air Staff |
| AC/S | Assistant Chief, Staff |
| ACMC | Air Corps Maintenance Command |
| ACTC | Air Corps Training Centers |
| ACTTC | Air Corps Technical Training Command |
| AD | Air Depot |
| A-Day | Activation Day |
| ADG | Air Depot Group |
| ADTS | Air Depot Training Station |
| AF | Air Force |
| AFACT | Assistant Chief, Air Staff, Training |
| AFASC | Air Service Command, Hq AAF |
| AFCC | Air Force Combat Command |
| AFDAE | Air Forces, Air Engineer |
| AFDMR | Air Forces, Director Military Requirements |
| AFDPU | Air Forces, Program Planning |
| AFIHI | Air Forces, Intelligence, Historical Section |
| AFRBS | Air Forces, Director Base Services |
| AFRIT | Air Forces, Director of Individual Training |
| AFSC | Air Force Service Command |

| | |
|---|---|
| AFTTC | Air Forces, Technical Training Command |
| AG | Adjutant General |
| AGCT | Army General Classification Test |
| AGD | Adjutant General's Department |
| AGF | Army Ground Forces |
| AGO | Adjutant General's Office |
| AMC | Air Materiel Command |
| AO | Airdrome Office |
| AR | Army Regulation |
| ARES | Aircraft Radio Engineering School |
| AS | Air Service |
| ASAC | Air Service Area Command |
| ASC | Air Service Command |
| ASCCO | ASC, Control Office |
| ASCPD | ASC, Personnel and Training Division |
| ASCPM | ASC, Military Personnel Section |
| ASCPM2B | ASC, Military Personnel Section, Unit 2B |
| ASCPP | ASC, Program Planning Office |
| ASCPRD | ASC, Personnel Replacement Depot |
| ASCPT | ASC, Military Training Section |
| ASCSUP | ASC, Supply Division |
| ASCTC | ASC, Training Center |
| ASF | Army Service Forces |
| asgd | Assigned |
| Assoc | Associated |
| A(A)SWAAF | Arms and Services with Army Air Forces |
| ATSC | Air Technical Service Command |
| AUS | Army of the United States |
| Avn | Aviation |
| AW | Articles of War |
| AWOL | Absent without Leave |
| B&G | Buildings and Grounds |
| B/L | Basic Letter (or Bill of Lading) |
| Bn | Battalion |
| BOQ | Bachelor Officers' Quarters |
| BU | Base Unit |
| CAC | Chief of Air Corps |
| C/AS | Chief, Air Staff |
| CBI | China, Burma, India Theater |
| CCC | Civilian Conservation Corps |
| CE | Corps of Engineers |
| CG | Commanding General |
| Chem | Chemical |

| | |
|---|---|
| Cir | Circular |
| Cml | Chemical |
| Cmt | Comment |
| CMTC | Citizens' Military Training Corps |
| Co | Company |
| CO | Commanding Officer |
| Com | Command |
| Comd | Command(ing) |
| Comm | Communications |
| Comp | Complement |
| C/S | Chief of Staff |
| Ctr | Center |
| CTU | Combat Training Unit |
| CWS | Chemical Warfare Service |
| D+2 | Day of Attack, plus 2 |
| DAR | Daily Activity Report |
| DC/AS | Deputy Chief, Air Staff |
| Dep | Deputy |
| Det | Detachment |
| Dir | Director |
| Div | Division |
| DP | Distribution Point (or By Direction of the President) |
| DS | Detached Service |
| DSO | Depot Supply Officer |
| EAUTC | Engineer Aviation Unit Training Center |
| EM | Enlisted Men |
| Eng(r) | Engineer |
| ETO | European Theater of Operations |
| FAD | Fairfield Air Depot |
| FASC | Fairfield Air Service Command |
| FATSC | Fairfield Air Technical Command |
| Fi | Finance |
| FM | Field Manual |
| FO | Field Order |
| FS | Field Services |
| FSS | Field Service Section |
| Ft | Fort |
| FY | Fiscal Year |
| G-1 | General Staff, Personnel |
| G-2 | General Staff, Intelligence |
| G-3 | General Staff, Training and Operations |
| G-4 | General Staff, Supply |

| | |
|---|---|
| G-5 | General Staff, Plans |
| GHQ | General Headquarters |
| GO | General Order |
| Gp | Group |
| HASC | Headquarters, Air Service Command, Washington, D.C. |
| HF | High Frequency |
| Hq | Headquarters |
| Hvy | Heavy |
| I&M | Installation and Maintenance |
| Ind | Indorsement |
| Insp | Inspection |
| ITT | Individual Technical Training |
| JCS | Joint Chiefs of Staff |
| jd | Joined |
| LCT's | Landing Craft Tanks |
| Lt | Light (or, of course, Lieutenant) |
| Ltr | Letter |
| M | Mimeographed |
| Maint | Maintenance |
| MASC | Middletown Air Service Command |
| MATSC | Middletown Air Technical Service Command |
| MC | Materiel Center (or Command) |
| Med | Medical |
| Mil | Military |
| Mil'd | Militarized |
| MM | Medium Maintenance |
| MM&D | Materiel, Maintenance, and Distribution |
| MOASC | Mobile Air Service Command |
| MOATSC | Mobile Air Technical Service Command |
| MOS | Military Occupational Specialty |
| MP | Military Police |
| MPTC | Military Police Training Center |
| MR | Military Requirements |
| MR-M-AF | Military Requirements, Mimeographed, Air Forces |
| MRU | Mobile Repair Unit |
| M&S | Materiel and Services |
| MTO | Mediterranean Theater of Operations |
| MU | Mobile Unit |
| NCO or Non-com | Non-Commissioned Officer |
| NYASPAC | New York Air Service Port Air Command |

| | |
|---|---|
| O | Officer |
| O-Day | Organization Day |
| OA&S | Other Arms and Services |
| OASC | Ogden Air Service Command |
| OATSC | Ogden Air Technical Service Command |
| OCAC | Office Chief of Air Corps |
| OCASC | Oklahoma City Air Service Command |
| OCATSC | Olahoma City Air Technical Service Command |
| OC&R | Operations, Commitments, and Requirements |
| OCS | Officers' Candidate School |
| OEL | Organizational Equipment List |
| Off | Office(r) |
| OIC | Officer in Charge |
| OPD | Operations Division |
| OQMG | Office Quartermaster General |
| Ord | Ordnance |
| OSW | Office Secretary of War |
| OTS | Officers' Training School |
| OTU | Operational Training Unit |
| OUTC | Ordnance Unit Training Center |
| PACMC | Provisional Air Corps Maintenance Command |
| P&BS | ATSC, Personnel and Base Services Division |
| Pers | Personnel |
| Plat | Platoon |
| PMP | Protective Mobilization Plan |
| POD | Port of Debarkation |
| POE | Port of Embarkation |
| POM | Preparation for Overseas Movement |
| PQ | Code name for first B-29 project |
| Prov | Provisional |
| P&T | Personnel and Training Division |
| Q | Quartermaster |
| (Q) | Type of Ordnance Company, transferred from Quartermaster |
| QM | Quartermaster |
| RAF | Royal Air Force of Great Britain |
| RASC | Rome Air Service Command |
| RATSC | Rome Air Technical Service Command |
| Recon | Reconnaissance |
| Refrig | Refrigeration |
| Reg | Regulation |
| ROTC | Reserve Officers' Training Corps |

| | |
|---|---|
| Rep | Repair |
| R&R | Routing and Record Sheet |
| RTU | Replacement Training Unit |
| S-1 | Staff, Personnel |
| S-2 | Staff, Intelligence |
| S-3 | Staff, Training and Operations |
| S-4 | Staff, Supply |
| SAAD | San Antonio Air Depot |
| SAASC | San Antonio Air Service Command |
| SAATSC | San Antonio Air Technical Service Command |
| SASC | Sacramento Air Service Command |
| SATSC | Sacramento Air Technical Service Command |
| SBASC | San Bernardino Air Service Command |
| SBATSC | San Bernardino Air Technical Service Command |
| SC | Signal Corps (or Statistical Control) |
| Sec | Section (or Security) |
| Secur | Security |
| Sep | Separate (or colored) |
| Serv | Service |
| Sess | Session |
| SG | Service Group |
| SG (Sp) | Service Group (Special) |
| SGTC | Service Group Training Center |
| Sig | Signal |
| S&M | Supply and Maintenance |
| SO | Special Order |
| SOP | Standing Operating Procedure |
| SOS | Services of Supply |
| SPASC | Spokane Air Service Command |
| SPATSC | Spokane Air Technical Service Command |
| Sq(n) | Squadron |
| SSN | Specialty Serial Number |
| Stat | Statistical (or Statutes) |
| Subsis | Subsistence |
| Sup | Supply |
| SW | Secretary of War |
| T | Teletype |
| T/A | Table of Allotment (or Table of Allowances) |
| TAD | Training Aids Division |
| TAG | The Adjutant General |
| TAT | To Accompany Troops |
| T/BA | Table of Basic Allotments (Allowances) |

| | |
|---|---|
| TC | Transportation Corps |
| TD | Temporary Duty |
| T/E | Table of Equipment |
| Tech | Technical |
| TF | Training Film |
| TM | Technical Manual |
| TO | Technical Order |
| T/O | Table of Organization (or Theater of Operations) |
| T&O | Training and Operations |
| T/O&E | Table of Organization and Equipment |
| T of O | Theater of Operations |
| Trg | Training |
| Trk | Truck |
| Trng | Training |
| TSAGD | ATSC, Adjutant General's Department |
| TSAIR | ATSC, Air Inspector's Office |
| TSBFO | ATSC, Budget and Fiscal Office |
| TSCHI | ATSC, Historical Office |
| TSCMC | ATSC, Management Control |
| TSCON | ATSC, Construction Office |
| TSCSU | ATSC, Statistical Control |
| TSDEP | ATSC, Deputy Commander |
| TSHIS | ATSC, Historical Office |
| TSMAE | ATSC, Associated Equipment Section, Maintenance Division |
| TSMCO | ATSC, Maintenance Control Office |
| TSPCO | ATSC, Control Section, Personnel and Base Services Division |
| TSPCP | ATSC, Civilian Personnel Section |
| TSPER | ATSC, Personnel and Base Services Division |
| TSPMP | ATSC, Military Personnel Section |
| TSPTR | ATSC, Military Training Section |
| TSSDL | ATSC, Disposal Section |
| TSSOV | ATSC, Overseas Supply |
| TSSUP | ATSC, Supply Division |
| TWX | Teletypewriter Exchange |
| USN | United States Navy |
| USO | United Service Organizations |
| VHB | Very Heavy Bombardment |
| VHF | Very High Frequency |
| VLR | Very Long Range |

| | |
|---|---|
| VOCO | Verbal Orders of the Commanding Officer (usually for leave of absence) |
| VS | Very Satisfactory |
| WAD | Wellston Air Depot (Warner Robins) |
| WAR | Weekly Activity Report |
| WD | War Department |
| WDGS | War Department General Staff |
| WEMA | Welfare, Enlisted Men, Army |
| WO | Warrant Officer (or Warning Order) |
| WRAAD | Warner Robins Army Air Depot |
| WRAD | Warner Robins Air Depot |
| WRASC | Warner Robins Air Service Command |
| WRATSC | Warner Robins Air Technical Service Command |
| ZI | Zone of Interior |

# INDEX

Bei Fragen zur Produktsicherheit wenden Sie sich bitte an:
If you have any questions regarding product safety,
please contact:

Walter de Gruyter GmbH
Genthiner Straße 13
10785 Berlin
productsafety@degruyterbrill.com